MEMS Product Engineering

Dirk Ortloff • Thilo Schmidt • Kai Hahn
Tomasz Bieniek • Grzegorz Janczyk • Rainer Brück

MEMS Product Engineering

Handling the Diversity of an Emerging Technology. Best Practices for Cooperative Development

Springer

Dirk Ortloff
Process Relations GmbH
Dortmund, Germany

Kai Hahn
Rainer Brück
Naturwissenschaftlich-Technische Fakultät,
Lehrstuhl Mikrosystementwurf
Universität Siegen
Siegen, Germany

Grzegorz Janczyk
Department of Integrated Circuits
and Systems
Instytut Technologii Elektronowej
Warsaw, Poland

Thilo Schmidt
Elmos Semiconductor AG
Dortmund, Germany

Tomasz Bieniek
Division of Silicon Microsystem
and Nanostructure Technology
Instytut Technologii Elektronowej
Warsaw, Poland

ISBN 978-3-7091-0705-8 ISBN 978-3-7091-0706-5 (eBook)
DOI 10.1007/978-3-7091-0706-5
Springer Wien Heidelberg New York Dordrecht London

Library of Congress Control Number: 2013951431

Printed on acid-free paper

Springer is part of Springer Science+Business Media (www.springer.com)

Foreword

I love it when plans come together. That's how I felt when Dirk Ortloff invited me to write this foreword. I see the genesis for this book on MEMS product engineering as a sign of the maturation of the MEMS industry as well as a little bit of serendipity. For some time now, and independent from the efforts resulting in this book, MEMS Industry Group (MIG) members have been tackling the issue of new MEMS product development. That topic was also the theme of MIG's annual technical conference, Member to Member (M2M) Forum. MIG surveyed the industry, interviewed thought leaders, and assembled an impressive body of knowledge to enable our members to address this challenge to MEMS commercialization. Amazingly (or perhaps I should say obviously) MIG's efforts lead to a similar outcome as presented by the authors of this book. This book has a slightly broader scope, covering aspects other than technical. It goes into more detail on the topic of MEMS product engineering and provides more hands-on business processes and tools.

One of the most important products of M2M Forum is the Technology Development Process (TDP) template, which our members developed under the leadership of David DiPaola of DiPaola Consulting, Peter Himes of Silex Microsystems, Tina Lamers of Avago, Valerie Marty of Hewlett-Packard, and Jason Tauscher of Microvision. The MIG TDP template provides a process framework for communicating expectations and facilitating alignment among the numerous partners involved in the process. It sets forth high-level objectives and outlines roles and responsibilities for all those involved in the product development process. The MIG TDP template also gives detailed examples, deliverables, and a gate-based decision-making framework to guide users as they adapt it to their needs. This framework can be supplemented and detailed by the contents of this book.

Our TDP template and the corresponding whitepapers are exemplary of how we do things at MIG because it's open source and stems from a collaborative process involving representatives of the entire MEMS supply chain—from materials suppliers, equipment vendors, and foundries to device manufacturers and end users/OEMs. It's a great example of a good thing done right. And the materials are free! They are available for anyone to download on the MIG web site—making them our gift to the MEMS industry. As long as you remember to say thank you, that is.

Through this book, the authors have helped to focus the conversation about MEMS product engineering and have documented more detailed processes, more best practices, and more tools—but things shouldn't stop here. MIG invites you to join the discussion about MEMS commercialization and how the MIG TDP template and the contents of this book can be a good starting point for your company, for your idea, and for your product. How have you tweaked it? How has it worked for you? How have you learned?

Central to MIG's mission is to provide a neutral forum for examining the technical challenges to MEMS commercialization for the entire supply chain. Part of achieving that goal is to extricate MEMS from the legacies of the semiconductor business. While semiconductor technology is part of our foundation, it is through the iterative process of adapting it to MEMS that we achieve innovation and success. Books like this one and products like the MIG TDP can be tools in your toolbox to help you achieve that success.

But it's important to remember that working with MEMS is a humbling process. And honestly, when it comes to MEMS technology development there is no one perfect "recipe" here. There is nothing like the one and only right business process. By learning together instead of independently, by applying best practices as documented in this MIG TDP template materials and this book, we can speed time to market and overcome those challenges faster. So read this book, adapt the practices, and process to your needs but then put it down and call a colleague; go to a workshop (hopefully an MIG one!); listen to a webinar; visit the MIG resource library that has a plethora of information on this topic; and continue the process of creating a working model and a collaborative engineering process in your own workplace for overcoming your own product development challenges and designing for manufacturability. You will make mistakes. You will take one step forward and two steps back, but through that process of collaboration and healthy engagement with your business partners, you will be able to take giant leaps beyond what you would have achieved independently-while avoiding common pitfalls through learning from documented best practices. And in the end, you will have a more robust, market-ready product. The business processes in the MIG TDP template and in this book enable you to continuously adapt to the "real world" of business, even though your product engineering process may not change much at its core. If your decision-making process foundation is solid, your products will be, too.

So read on with confidence, knowing that while MEMS may not be for the faint at heart, that there are people, resources, and tools out there to help you. With double-digit annual growth, the rewards of following best product engineering principles for new MEMS product development are great, and there is definitely room for you to jump in and join us, to take advantage of this growing market opportunity.

Pittsburgh, PA Karen Lightman
March 2013

Preface

The creation of a novel technological product always starts with an idea, a collection of more or less imprecise conceptions of how it should work, what it should look like, and why the market would appreciate it and thus might turn it into an economic success. This idea is usually driven by a customer. But customer does not necessarily mean the end user or consumer. The supply chain in markets for technological products is typically quite long. So in this book the customer could be a different department of a single legal entity or a technological integrator like an electronics manufacturer. In any of these scenarios the customer is the only one who has all the required knowledge concerning functionality and the constraints of the application area. This makes the customer the natural instance to be in charge of leading and guiding the process that turns his/her idea into a marketable product.

Between the idea of the customer and millions of copies of the physical product or device ready for sale and use, a usually lengthy period of time goes by. This period is characterized by creativity, complex decision processes, setbacks, and lots of money spent. In the automotive industry in Europe, for example, it takes about 5 years from the idea of a new car to start of production. In consumer electronics, although a market with an extreme greed for innovation, this process still takes 6 months to 1 year. The increasing dynamics of modern markets for technological products and devices requires this process to be as short as possible, free of errors and yielding cost-effective, safe, secure, and optimized products. This is possible only if the path from the idea to the final product is accompanied by a systematic, controlled, and safe product creation/product engineering process.

Product Engineering is a term that denotes this sort of systematic processes guiding and controlling product creation from the very idea to the final devices ready for sale. The effort involved in a specific product engineering process is highly sensitive to the functional complexity of the intended product, the involved collaboration partners along the value chain, and the technological complexity of the required fabrication facilities.

Two aspects of modern high-tech products take influence on the composition and the complexity of the product engineering process. The first aspect is that the fabrication of one product frequently makes use of several very diverse fabrication methods. Today, in many cases not all of these methods are available at a production facility. In this case one product requires the cooperation of several fabrication facilities, generally available only from different specialized companies. The second aspect is a direct consequence of this requirement for cooperative fabrication: Each of the partners involved in the fabrication effort has dedicated knowledge concerning their specific fabrication capabilities. But no one of the partners has sufficient knowledge of the overall requirements of the usually complex product. This role has to be taken and driven by the customer giving the customer a key role in the product engineering process.

Distributed, customer-oriented product engineering is the process required for innovative complex technological products. This type of products is found in all areas that are commonly regarded as "high tech" today. This covers, for example, automotive and aerospace industries, information technology, and more and more new fields like biomedical or micro- and nanotechnology. The field of micro- and nanotechnologies (MNT) is a particularly well-suited example for an industry where customer-oriented, distributed product engineering processes are needed. MNT needs to cope with the increasing demand for shorter product engineering cycles while at the same time production makes use of the leading edge of the fast-developing fabrication capabilities. The origin of MNT fabrication is in microelectronics. There, for the last 50 years, a centralized product engineering scenario has reached a very high level of maturity. In this "classical" scenario one company, an Integrated Device Manufacturer (IDM), was in charge of the whole product engineering process from the idea to the shipped products. Two strands of progression in the last 15 years made this scenario at least partially obsolete. The first one is the strict adherence of the semiconductor industry to the famous Moore's Law driving the feature sizes of electronics devices continually deeper into the nanometer range. This calls for extremely sophisticated fabrication facilities that demand investments that are frequently beyond the capabilities of classical IDM microelectronics suppliers. The business model of pure play foundries at the rear end of the product engineering process has becoming more and more popular. In that model pure play foundries perform only high-volume fabrication using one manufacturing technology for diverse customers. The fabless (fablight) design houses at the front end turn product ideas into a set of appropriate inputs (specs, mask designs, etc.) for volume fabrication at the pure play foundries. This is the first step towards a distributed product engineering scenario. The second relevant progression is the increasing demand for heterogeneous integration. Many of today's technological products offer functionality that requires sensors, electronics, and actors to be integrated into one single system. Products of this type are generally addressed as *micro electro mechanical systems (MEMS)*. The issue here is that sensors, actors, and electronics are usually based on different fabrication technologies. This leads to product engineering scenarios where fabrication is distributed among different companies, each one specialized in either sensors or actors or electronics fabrication. With assembly, packaging, and test of the products yet performed by other specialized companies, product engineering efforts are distributed among a large group of different parties.

Research and development efforts performed in the authors' companies and institutes during the last decade were dedicated to the problem of finding a generic methodology for customer-oriented, distributed product engineering of MNT products. The research effort culminated in the cooperative project Customer-oriented product engineering of micro and nano devices (CORONA). The project was funded by the European Commission in the 7th research framework programme (grant agreement CP-FP 213969-2). The project was conducted from 2008 to 2011 by 9 partners from 5 European countries as indicated in the acknowledgments. The goals of the project were threefold. First was defining, implementing, and optimizing a generic methodology for distributed, customer-oriented product engineering in MNT, based on established baseline methodologies from other fields and on first methodological results obtained from previous research by the authors. The second goal of the project was developing software tools to support the execution of the methodology. The third goal was to prove the applicability of the methodology by real-life demonstrators. This book is the central result of the methodology work done in the CORONA project.

This book is intended for every person who takes an active role in either doing product engineering or teaching how to do it. This includes project and product management staff or program management offices in companies practicing innovation projects, every person active in doing innovation, as well as professors and students in engineering and management.

The book covers customer-oriented, distributed product engineering in the area of MNT. The methodology, however, is generic in wide areas and is also applicable to high-tech industries other than MNT. With some domain-specific modifications the methodology even appears useful as a baseline in other industrial areas where distributed and/or customer-oriented engineering is or becomes common.

Guide to the Book

This book addresses all persons interested in product engineering no matter what their professional background is. However, the book presents product engineering in a very specific setting. It thus consists of an introductory part that attempts to provide all the necessary background knowledge for everyone to be able to understand and appreciate the further chapters that treat the methodology and its application. In detail the book is organized as follows:

Part I is intended to pave the path to the book for every interested reader, especially for those who are not familiar with the field of MNT. It introduces the application area that has been selected to develop the product engineering methodology presented in the further chapters.

- *Chapter 1* gives an introduction to the MNT industry, market, and supply chain. It shows the specific properties of the market, shows, in a comprehensive manner, how MNT product engineering basically works, and introduces the challenges and goals of a formal product engineering methodology for this area. In this manner a lean and comprehensive introduction to all areas touched in the remaining chapters is given.

- *Chapter 2* is dedicated to the term "product engineering" as it is understood by the authors. It gives definitions of terms and describes the essentials of how to perform innovation and the engineering of new products. The methodology that will be described in detail in Chap. 5 is derived from several baseline methods. Some of them like Stage-Gate® or PRINCE2® are common and have been in use in other application areas than MNT for quite some time, other ones like the T^2M methodology have been introduced by the authors. The chapter gives a very brief introduction to various possible baseline methods and to the original literature that covers these methods in all required detail. It furthermore gives reasons why specific baseline methods have been selected as foundations for the new methodology presented in Chap. 5.
- *Chapter 3* covers micro and nano systems. This chapter is essential for readers with little or no background in this field. Micro- and nano systems have been used as the application area for which the methodology presented in this book has been optimized. To be able to fully understand and appreciate the specifics of the new methodology, a minimum understanding about MNT is necessary. The chapter gives a brief and easy-to-understand introduction to micro- and nano systems, how they work, how they are produced, and who is involved in creating this sort of products. The interested reader is guided to more elaborate texts on this topic in a comprehensive reference to the relevant literature.

Part II of the book covers MEMS product engineering in detail from three different perspectives. It is the central part of the book that introduces the reader to all relevant details of MEMS engineering processes and the formal methodology on how to conduct them, which the authors developed and implemented in their research work.

- *Chapter 4* addresses MEMS product engineering from the teams' point of view. It introduces the reader to the various tasks that have to be performed during a typical product engineering process. It covers the aspects of device design, technology design, and quality management. It shows why MEMS require a distributed product engineering approach and why it is essential to put the customer in charge of this process. It furthermore gives examples of how and why software support is utilized when conducting MEMS product engineering projects.
- *Chapter 5* gives a thorough introduction to the product engineering methodology developed by the authors during the CORONA project. After an overview of the method and how it is composed from various building blocks it dives deep into all the details that are necessary to understand the methodology and to use it as a blueprint to set up an arbitrary product engineering project in the area of MNT.
- *Chapter 6* again changes the point of view and takes the perspective of the practical application of the methodology. It gives an overview of where and why it is required to have extensive tool support to perform product engineering

projects based on the methodology. It describes the main ingredients that are necessary to build up an appropriate software infrastructure for MEMS product engineering. However, it does not give recommendations to specific software packages to be used. This decision is intentionally left to the reader and user of the methodology as it is usually dependent on company policies and given software infrastructures.

Part III of the book is dedicated to the industrial product engineering practice. It only consists of *Chap. 7* that gives an overview of a practical product engineering endeavor that has been performed in a real-life distributed product engineering project by the authors. This project has served as a demonstrator within the CORONA project. It gives a very clear indication of the usefulness of distributed engineering for real-life MNT products with the customer in charge of the product engineering process. In this manner it can serve at the same time as evidence for the efficiency and effectiveness of the methodology presented in Chap. 5 and as a practical guide on how to set up a product engineering endeavor based on this methodology.

Sect. 8 finally complements Chap. 5 by giving more details of the project management sequences, the steps of the processes and their dedicated inputs, and outputs required when conducting a MEMS product engineering project.

Dortmund, Germany — Dirk Ortloff, Thilo Schmidt
Siegen, Germany — Kai Hahn, Rainer Brück
Warsaw, Poland — Tomasz Bieniek, Grzegorz Janczyk
June 2013

Acknowledgments

The developments of the methodology and this book have been carried out and supported by an international multisite research project named CORONA—funded by the European Commission under the 7th Framework programme under the agreement CP-FP 213969-2. We are grateful and appreciate the input from all project partners, namely

Coventor Sarl
Dr. Gerold Schröpfer
Villebon-sur-Yvette, France

ELMOS Central IT Services GmbH
Dr. Ralf Montino
Dortmund, Germany

Instytut Technologii Elektronowej
Dr. Tomasz Bieniek
Warsaw, Poland

IVAM e.V. Coordinator
Inga Goltermann M.A.
Dortmund, Germany

Process Relations GmbH
Dr. Jens Popp
Dortmund, Germany

Theon Sensors S.A.
Emmanuel Patsios
Athens, Greece

Universität Siegen
Prof. Dr. Rainer Brück
Siegen, Germany

University of Cambridge
Dr. Andrew Flewitt
Cambridge, United Kingdom

X-FAB Semiconductor Foundries AG
Dr. Gisbert Hölzer
Erfurt, Germany

Furthermore the authors of this book would like to acknowledge all the coworkers who have contributed to the work and participated in numerous fruitful technical discussions. From Division of Silicon Microsystem and Nanostructure Technology and Department of Integrated Circuits and Systems of the Institute of Electron Technology, Poland (ITE), we would like to acknowledge and thank especially Piotr Grabiec, Pawel Janus, Magdalena Ekwinska, Krzysztof Domanski, Andrzej Sierakowski, Dariusz Szmigiel, Stanislaw Kalicinski, Jerzy Wasowski and the whole ITE technical staff involved for creative cooperation. Furthermore we would like to say a special thanks to the contributions from Jens Popp, Felix Rotthowe, Michael Rautert, Thanasis Kollias, and Uwe Schwarz.

A special thank you goes to the MEMS Industry Group (MIG) and the group's executive director Karen Lightman for providing the foreword to this book. The open collaborative working style provided for allowing the integration of parts of their Technology Development Process (TDP) template [1] into the framework for this book. The TDP template is one of the most important products of M2M Forum 2012, which the MIG members developed under the leadership of David DiPaola of DiPaola Consulting, Peter Himes of Silex Microsystems, Tina Lamers of Avago, Valerie Marty of Hewlett-Packard, and Jason Tauscher of Microvision. The MIG TDP template provides a process framework for communicating expectations and facilitating alignment among the numerous partners involved in the process. It sets forth high-level objectives and outlines roles and responsibilities for all those involved in the product development process. The MIG TDP template also gives detailed examples, deliverables, and a gate-based decision-making framework to guide users as they adapt it to their needs. The collaboration in merging the efforts and jointly improving the approaches and materials has been very fruitful, and we are looking forward to a continued collaboration and interaction with the user community.

Reference

1. MEMSIndustryGroup: Technology development process template. http://www.memsindustrygroup.org/i4a/doclibrary/getfile.cfm?doc_id=373 (2012). Accessed 10 March 2013

Contents

Part II MEMS Product Engineering Aspects and Methods

Acronyms

BOM	Bill of materials
CE	Concurrent engineering
CEO	Chief executive officer
CIO	Chief information officer
CM	Configuration management
CMMI	Capability maturity model integrated
CR	Change request
CTO	Chief technology officer
DFM	Design for manufacturability
DfSS	Design for Six Sigma
DICOV	Define, identify, characterize, optimize, validate
DMADV	Define, measure, analyze, design, verify
DoE	Design of experiments
DRC	Design rule check
EDA	Electronic design automation
FEM	Finite element modeling
FMEA	Failure mode and effects analysis
IDM	Integrated device manufacturer
IDOV	Identify, define, optimize, validate
IP	Intellectual property
IPPD	Integrated product and process development
ITIL	IT infrastructure library
KPI	Key performance indicator
LVS	Layout versus schematic
MIG	MEMS industry group
MEMS	Micro electro-mechanical systems
MES	Manufacturing execution system
MNT	Micro- and nano technology
MPW	Multi project wafer
M&S	Marketing & sales
OEM	Original equipment manufacturer

PDES	Process development execution system
PE	Product engineering
PEF	Product engineering framework
PEM	Product engineering method
PM	Project management
POC	Prove of concept
QA	Quality assurance
QFD	Quality function deployment
QMS	Quality management system
R&D	Research and development
RE	Requirements engineering
RMI	Remote method invocation
RS	Requirements specification
RTL	Register transfer level
TDP	Technology development process
TTM	Time to market
VHDL	Very high speed integrated circuit hardware description language
VLSI	Very large scale integration
VOC	Voice of the customer
WBS	Work breakdown structure
XP	eXtreme programming

List of Figures

List of Tables

Part I
General Introductions

This first part of the book is intended as an introduction to the topics relevant to MEMS Product Engineering as covered by this book.

Chapter 1 gives a general overview and explains the goals of the book. It also introduces the challenges and specifics of the MEMS market providing the background for the development of the methodology presented later in this book.

Chapter 2 describes the various aspects of Product Engineering as perceived by the authors and introduces generally available and established methods.

The first part is concluded by Chap. 3 about the MNT industry—the area the Product Engineering methodology is targeted at. This chapter gives a brief summary about what MEMS are, how they are made, and who is involved in making them.

Chapter 1
Introduction

Abstract Over the last years MEMS have become a major driving force in the automotive and consumer electronics markets. Small and inexpensive acceleration sensors, cameras, microphones, micro-mirrors, touch-panels, and other products of the MEMS industry are by now the key components of almost every new and innovative electronic appliance. A fast and effective Product Engineering approach is needed to be competitive in such fast moving markets. This chapter introduces the market constraints and limitations and defines the goals for collaborative, customer-oriented Product Engineering.

1.1 The Goal

During the last years there has been a massive increase of consumer products based on micro- and nanotechnology (MNT), like cell phones and home entertainment appliances. New and manifold application areas, short product cycles, and tight competition are challenging product engineering in this field. The lack of a common technology platform like CMOS in the semiconductor industry [8] led to a mostly technology-driven product engineering approach: After the development of new technological capabilities, a survey is carried out for device concepts and markets that profit from the new technology [11]. Emerging MNT business models like the *fabless design house* or the *pure play foundry* that are already commonplace in the area of microelectronics call for new design approaches that focus on market needs and collaborative design. The methodology proposed in this book therefore regards the customer, being the representative of the market and typically not being a consumer, as the main driving force behind the product engineering process.

Regardless of this intended shift towards a more market-driven product engineering approach, the huge impact of manufacturing technology on product engineering is still present. The engineering of micro- and nanotech devices based on semiconductor manufacturing processes consists of the structural design as well

D. Ortloff et al., *MEMS Product Engineering*, DOI 10.1007/978-3-7091-0706-5_1,

as the definition of the manufacturing process flow. The approach is characterized by application-specific fabrication flows, i.e., fabrication processes (built up from a large variety of process steps and materials) depending on the intended product. Recent market reports state that the so-called MEMS law ("One product, one process!") still holds [19]. Nearly every new product requires an individually tailored manufacturing process to match the constraints of the system design and the application area. Micro-mechanical devices need other material properties and geometries than micro-optical devices or micro-fluidic devices. A sensor in an automotive setting has to qualify for other constraints than a similar sensor in a mobile phone. Consequently, the design and development of an application-specific manufacturing process is in most cases still an integral part of MEMS product engineering. First steps towards technology platforms suitable to classes of devices are made but still require adaptation to the specific device constraints. Additionally these platforms are either provided by foundries typically being quite rigid concerning adaptations or changes in the flow or by Integrated Device Manufacturers (IDM). A new progression are the technology platforms offered even by fabless MEMS manufacturers. The most prominent example currently being pushed into the market is the InvenSense NF-Shuttle [1] offering a so-called Nasiri Fabrication platform based on standard CMOS processes and single-crystal SOI-based MEMS.

Despite all of these initiatives to, at least partly, standardize and speed up the engineering cycle and the Time to Market (TTM), designing a MEMS manufacturing process and device is still a very tedious and lengthy endeavor. This task still depends mostly on physical experiments, prototype runs, and the individual experience of the involved engineers. Careful information and knowledge management can speed up this process by utilizing experiences and experimental results gained from previous projects. An important prerequisite for this is that the actual data gained during experiments and prototype runs is collected holistically, categorized in an accessible, comprehensive way and that the complete context and the complete timeline of the engineering process remain available. Due to the multidisciplinary nature of MNT development teams the required infrastructure needs to provide a multi-dimensional view to matters to be able to align the different perspectives and requirements different disciplines have.

This book introduces a comprehensive methodology for MEMS product engineering. The methodology can be used to guide engineering efforts from initial ideas to shipping products. The keys of the approach are the customer-centric business processes and the flexibility to involve as many collaboration partners as required along the MEMS value chain. Therefore the methodology allows to set up projects for success and to guide engineering efforts through all possible situations in a way as versatile and as efficient as possible. The method is supplemented by a set of software tools that integrate design (device and process), corporate knowledge management, virtual manufacturing, and data acquisition [6, 12, 13, 15, 16].

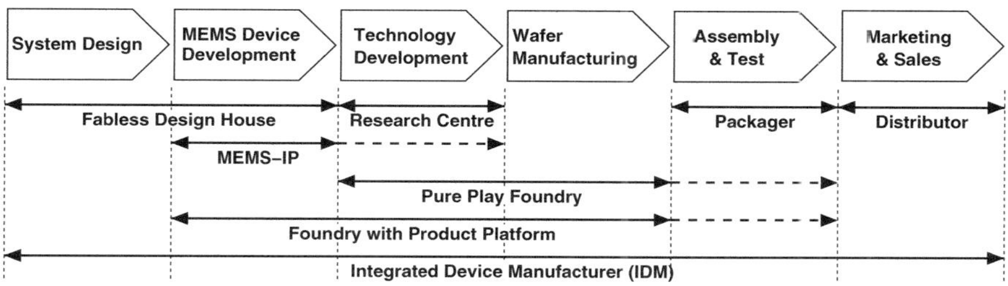

Fig. 1.1 MNT value chain

1.2 Challenges in MNT Product Engineering

Implied by the particular nature and collaboration structure of the MNT industry, e.g., split-up value chain, several characteristic challenges emerge for MNT product engineering. New and emerging application areas, short engineering cycles, and tight competition are the main challenges for product engineering in this field. Besides a relatively small number of "traditional" large micro- and nano electronics enterprizes, the MNT sector consists mostly of small- and medium-scale enterprizes (SME). These entities cannot and do not offer the complete engineering or value chain. Rather, they offer products, services, and knowledge for a specific area, like special fabrication techniques, packaging services, design services, IP-based products like calculation cores or pure process IP. Nontraditional business models (like the fabless design house or the pure play foundry) that are already common in the field of microelectronics require new product engineering approaches that focus on collaborative design and manufacturing due to the complexity of the value chain. Therefore collaboration along the value chain as depicted in Fig. 1.1 implies a major challenge to the engineering organizations in the MNT field.

Product innovation in the MNT industry is in most cases a collaborative effort of several independent companies across the value chain. A customer can contract the companies for the development of a specific product or part of a product (then using several partners to collaboratively engineer the product). This collaboration is accompanied by pitfalls arising from the distribution of tasks and knowledge as well as the interdependencies between the different product parts. In such an innovative setting, communication becomes a key success factor. The demand for short TTM requires deep interaction between the customer and the supplier, respectively, all involved partners. For the organizations involved in a product creation process, cooperation is necessary to create and produce a competitive product in a short time. Furthermore, due to the tight time constraints for developing successful products, customers require to be more tightly integrated into the group of collaborating partners.

As a consequence of the distributed, multi-site/multi-company creation of a product, no central controlling instance exists. The leading and coordinating position in the innovation process must be taken by the customer, since he/she has extensive knowledge about the product and its target market.

Besides the SME-predominance and the subsequent distributed innovation, the high technology-diversity is a main characteristic of the MNT sector. As stated above, many SMEs offer individual production steps. These technologies range from precision engineering, silicon micromachining to high aspect ratio techniques like LIGA. The customer must choose the appropriate technologies and their respective processes for the end product and the engineering process must incorporate and coordinate all involved stakeholders.

1.2.1 Distributed Engineering and Prevalent Business Models

As described above, the different companies in the MNT industry offer different services or products. Based on the offered services they can be classified as shown in Fig. 1.1. The services provided by different companies in the supply chain depicted in the figure cover the complete product engineering process from research to fabrication. The next sections show the main roles of companies involved in this scenario.

The pressure on innovation in high-tech and science led industries has been nicely summarized by IDC [2]. They basically argue that

- only 25 % of the innovation projects achieve product introduction to the market,
- of these 66 % fail to deliver their original expectations.
- 20 % of the innovation projects take too long and miss their market window.
- 35 % of all innovating companies experience runaway projects
- and a staggering number of 40 % of R&D experiments are repeated at least once.

In the light of these numbers and to give a moderate piece of help to reduce the huge investment that never makes it into the marketplace or all the waste produced due to insufficient engineering methodology and tool infrastructure this book has been written.

1.2.1.1 (Fabless) Design House

A (Fabless) Design House is a company offering design services for MNT devices. The company has extensive knowledge about the conception and design of a device, but only limited knowledge about fabrication. With the help of the customer, the design house develops a specification for a new MNT product. When the specification is fixed, it begins with the design process of the product or the device. The design house covers all aspects of behavioral and physical design of the new product. When the physical design is finished, the design house commits the fabrication data to a foundry, which manufactures the developed product. Often, a design house offers additional services to their customers, like ramping to mass production. Another facet of an entity following the fabless approach are fabless process IP companies. Rather than designing a product or fixed device, they develop process IP and design IP to be integrated into an overall device.

1.2.1.2 Foundry

A foundry is a company offering fabrication services of MNT devices. Unlike the design house the foundry has comprehensive knowledge about fabrication of micro- and nano devices, but only very limited design competence. The foundry receives the fabrication data, e.g., a mask layout, and fabricates the device at its facilities. The foundry is a contract manufacturer, for producing MNT devices in high quantities. Often foundries offer services for prototype, low- and mid-volume production and usually do not offer any product in the sense of readily produced devices in their own name.

In principle two types of foundries can be distinguished covering slightly different parts of the value chain in Fig. 1.1. The Pure Play Foundries focus on technology development and wafer manufacturing as described above. The other type of foundry offers product platforms on top. These are oftentimes technologies and device prototype modules for device categories like accelerometers, gyroscopes, etc.

1.2.1.3 Technology Development

Some organizations (companies or research institutes) offer new fabrication technologies and development of new fabrication processes. These organizations have very deep knowledge in processing techniques for micro- and nano-structures, sometimes in a particular process step like lithography or etching only. New process sequences for the fabrication of MNT devices are the main product of this type of organization. These process steps or sequences are offered on the market, can be licensed by a company and transferred to its own production line.

1.2.1.4 Assembly and Test House

Assembly and test houses can be distinguished into two types as well. The one type forms the backend of product manufacturing. As MNT devices often need special packaging and test procedures, fabricated dies are sent from the foundry to specialized assembly and test houses for packaging and testing.

Among the companies in the MNT sector, there are specialized companies for post processing of devices and for technology development. These can be considered a second type of assembly and test house. A MEMS-device usually consists of a non-electronic component and an electronic circuit, controlling or reading the non-electronic part. These two parts are often combined on one silicon die. On this die the electronics are fabricated first and in a post-processing process, the non-electronic parts are added to this die. The integration of the non-electronic component on the die is often done by specialized companies.

1.2.1.5 Integrated Device Manufacturer

An Integrated Device Manufacturer (IDM) offers all services covering the complete design- and fabrication phases of an MNT product. It has in-house design, fabrication, testing, and assembly facilities. An IDM can deliver a new product from a specification compiled with the customer. Normally the customer is only involved in the product specification phase. The development of the design and fabrication process is performed alone by the internal departments of the IDM. The customer is informed about the development and fabrication progress by regular status updates. At critical points in the development process, the customer has to approve to the progress made so far.

1.2.1.6 Additional Roles

A product engineering process does not only cover the development process of a product. Rather it covers the complete life cycle. As a consequence not only specialized companies from the MNT sector are involved in a product engineering process. Companies with other competencies complete the multifunctional team of the product engineering process. Among these companies are specialists for legal assessment and protection, marketing, distributors, and trading companies.

1.2.2 Contemporary Customer Involvement

To identify the degree of customer involvement in the MNT sector, the CORONA project conducted various interviews with companies along the MNT supply chain. The results, published in [14], show that the degree of customer involvement in contemporary innovation processes is rather limited and depends on the business model of the company and the product to be developed.

Independent of the business model and the product, the customer is mainly involved in the beginning of the product engineering process. The current approaches involve the customer mainly in the definition and specification of the new product. When a specification of the product has been completed by the innovation partners, the customer becomes passive. He/she is informed of the progress by regular status updates. These updates can be in the form of conferences, conference calls, prototypes, or by paper. In the steps following the specification the contribution of the customer is mainly the approval of milestones.

Another means for the customer to influence the progression after specification is a change request. If the customer discovers a potential risk, e.g., the product doesn't fit the requirements of the market any more and wants the specification to be changed, he/she can submit a change request. After this request, the customer and the partners in the innovation process discuss possible changes.

Besides this common procedures implemented by all interviewed companies, few companies offer the customer deeper interaction in the product innovation process. The major problem with this approach is that typically developments of MNT products take several years. During this time the legal as well as the business or market situation can change substantially such that developed device characteristics or even full devices are not commercially viable anymore. With this loose integration of the customer into the innovation process, valuable time adapting to market needs can be lost. Therefore new methods and tools for tighter, low effort integration of the customer as well as all the other stakeholders into the innovation process are imperative.

1.2.3 Technical Collaboration Challenges

To dive a little deeper into the Product Engineering (PE) challenges for MNT it is useful to try to find analogies to the semiconductor industry. In the last decades semiconductor progression has benefited from a clear separation between manufacturing and design. This separation was enabled by some characteristics of VLSI systems. Manhattan- or octagonal layout, fixed and well-known manufacturing processes, and a limited number of functional components allow a reduction from the original three-dimensional design to 2.5 dimensions (planar layout plus technology constraints delivered as geometric design rules). In combination with the Y-chart model of Gajski et al. [3, 4] this so-called Mead–Conway abstraction [10] forms the basis of nearly all design concepts and design automation in the field of microelectronics.

With the challenges offered by deep submicron, mixed-signal and also mixed-domain technologies that reduction is already problematic and for technologies like MEMS or NEMS (nano-electro mechanical systems) this approach is not directly applicable [9]. For MEMS the overall design flow is heavily influenced by manufacturing aspects. Changes at any design level have an impact on manufacturing and vice versa. Figure 1.2 illustrates at which design levels behavioral or fabrication-related aspects dominate. At the physical design level the device geometry meets the manufacturing process and the material properties. The three-dimensional geometry and the material properties that are mainly responsible for the functional behavior of the device depend mostly on the capabilities and constraints of the manufacturing process. That often leads to the development of application-specific processes ("MEMS law") which also means new design rules and technology files for virtually every design.

A short-term solution to handle the dependencies between behavioral design and fabrication can be a structured physical design flow as proposed by Brück, Hahn [5], and Wagener et al. [18]. The cycle model, as presented in Fig. 4.2, describes an

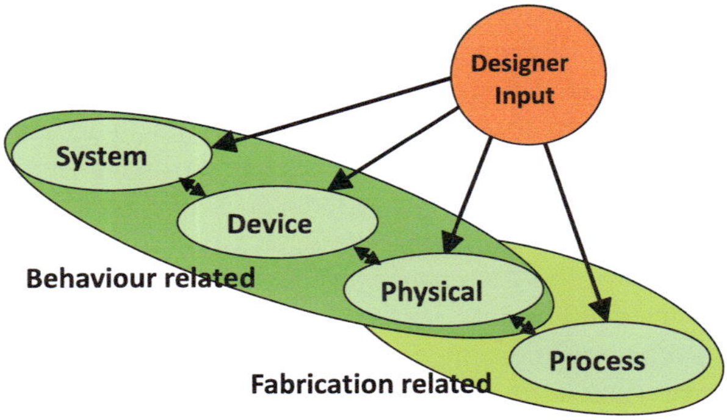

Fig. 1.2 Behavior and fabrication related design levels for Microsystems based on [17]

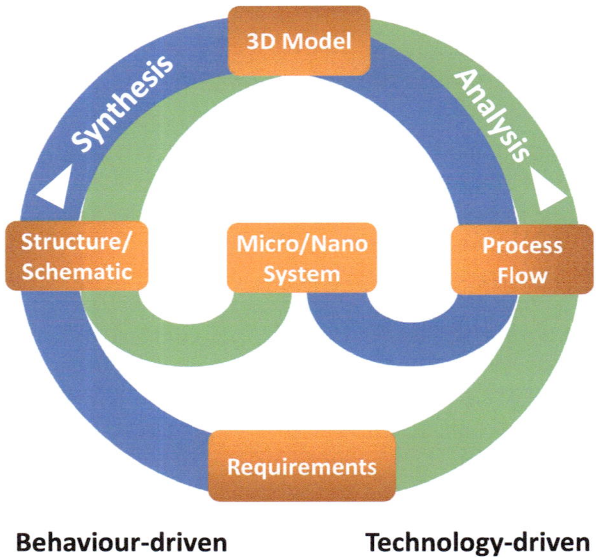

Fig. 1.3 Pretzel model for MEMS design and fabrication [7]

iterative design flow implementing design, validation, and redesign. Such a physical design flow is currently very common in the MEMS community. The disadvantage of this model is that it concentrates on the physical design aspects without including behavioral design activities. A more adequate model that explicitly takes into account behavioral as well as fabrication-related design is the so-called pretzel model [7] shown in Fig. 1.3. It consists of two interwoven flows: a top-down synthesis flow and a bottom-up analysis flow. The synthesis flow starts by creating a structural description (schematic) based on the system requirements. From this structural description a corresponding physical model of the device is derived. For MEMS this is typically a three-dimensional model. The three-dimensional model serves as a basis for an adequate manufacturing process that can be used to fabricate the MEMS device.

The bottom-up analysis flow works in the opposite direction. It starts by comparing the system requirements with the technological constraints, thus deciding if the selected manufacturing technology is adequate. Then a technology simulation or emulation is performed to check if the fabrication flow is able to actually create the 3D-model. To compare the 3D-model with the structural description behavioral simulations (e.g., FEM) are employed. The final step of the analysis flow is the comparison of the structural model with the device prototype (e.g., by test patterns). As a consequence the left part of the pretzel covers behavioral aspects whereas the right part is dominated by fabrication and technology aspects. The model does not imply when or in which order the design steps have to be performed. Similar to the Y-chart Model [3, 4], many different design flows can be devised.

At the moment the implementation of this approach is not supported adequately. Available design tools either cover the technology aspects (TCAD tools) like process assembly and simulation or there are specialized tools for the behavioral approach with support of, e.g., macro modeling, mask layout, and FEM simulation. There is still a lack of support at the interface between technology design and behavioral modeling. To close this gap this book proposes an approach based on a comprehensive knowledge base. All technology data is collected and prepared for the usage in behavioral as well as technological modeling. The required technology information is collected as a by-product from the physical design flow [15].

To enable such a comprehensive design flow, CAD tools have to be provided to collect and manage the fabrication relevant data and to extract the knowledge for the modeling. There is a need for behavioral design and modeling tools as well as comprehensive and sophisticated process management environment, which provides validated data for behavioral modeling.

Consequently, MNT product innovation also includes the development of appropriate manufacturing processes. MNT Product Engineering is therefore characterized by a large diversity of products, applications, and technologies and the lack of a common technology platform like CMOS [8, 11]. Therefore specifically adapted product engineering methodologies taking care of the diversity of products and technologies and the collaborative nature promise to be the key to the continued growth of the MNT industries.

References

1. Assaderaghi, F., Daneman, M., Lim, M.: Nf-shuttle: Standardizing the Fabless MEMS Industry. Invensens, Sunnyvale (2012). http://www.invensense.com/mems/gyro/documents/whitepapers/Yole-NFS-Whitepaper.pdf. Accessed 21 Aug 2013
2. Barkai, J.: Accelerating Science-led Innovation for Competitive Advantage. IDC Manufacturing Insights, Framingham (2012). http://www.winvale.com/pdfs/white_paper_module/IDC_Accelerating_Science-led_Innovation.pdf. Accessed 21 Aug 2013

3. Gajski, D., Kuhn, R.: Guest editors' introduction: new vlsi tools. Computer **16**(12), 11–14 (1983)
4. Gajski, D.D., Thomas, D.E.: Introduction to silicon compilation. In: Gajski, D.D. (ed.) Silicon Compilation, pp. 1–48. Addison-Wesley, Reading, MA (1988)
5. Hahn, K., Brück, R.: An approach to layout and process verification for microsystem physical design. Microsyst. Technol. **3**(2), 53–60 (1997)
6. Hahn, K., Schmidt, T., Ortloff, D., Popp, J., Wagener, A., Brück, R.: MEMS product engineering using fabrication process development tools. In: Micro- and Nanotechnology: Materials, Processes, Packaging, and Systems IV, vol. 7269. Photonics West 2008, Melbourne (2008)
7. Hahn, K., Wagener, A., Popp, J., Brück, R.: Process management and design for MEMS and microelectronics technologies. In: Proceedings of SPIE: International Symposium on Microelectronics, MEMS, and Nanotechnology 2003, Perth (2003). SPIE Perth 2003
8. van Heeren, H.: Appearance of a moore's law in MEMS? Trends affecting the MNT supply chain. In: Proceedings of SPIE 6186, MEMS, MOEMS, and Micromachining II, Strasbourg, France, pp. 61, 860D–61, 860D–12. SPIE, Bellingham (2006). doi:10.1117/12.662191. http://dx.doi.org/10.1117/12.662191
9. Hubbard, T.J., Antonsson, E.K.: Design of mems via efficient simulation of fabrication. In: Proceedings of the ASME Design Engineering Technical Conference and Computers in Engineering Conference, Irvine (1996). ASME96
10. Mead, C., Conway, L.: Introduction to VLSI Systems. Addison-Wesley, Reading, MA (1980)
11. Ortloff, D.: A product engineering method for MEMS ip developments. In: Proceedings of ICME: 5th CIRP International Seminar on Intelligent Computing in Manufacturing Engineering, Ischia (2006). CIRP ICME'06
12. Ortloff, D., Popp, J., Hahn, K., Schmidt, T., Bruck, R.: Tool support for microelectronic process development. In: 15th International Conference on Mixed Design of Integrated Circuits and Systems, pp. 467–472 (2008). MIXDES 2008
13. Ortloff, D., Popp, J., Schmidt, T., Brück, R.: Process development support environment: a tool suite to engineer manufacturing sequences. Int. J. Comput. Mater. Sci. Surf. Eng. **2**(3/4), 312–334 (2009)
14. Ortloff, D., Popp, J., Schmidt, T., Mielke, M.: A customer-driven approach to product engineering of micro and nano devices requirement analysis. In: Proceedings of the 13th International Conference on the Commercialization of Micro and Nano Systems, Copenhagen (2009). COMS 2009
15. Ortloff, D., Popp, J., Wagener, A., Hahn, K.: Breaking through the process development barriers. In: Proceedings of the 12th International Conference on the Commercialization of Micro and Nano Systems, Puerto Vallata (2008). COMS 2008
16. Popp, J., Ortloff, D., Schmidt, T., Hahn, K., Mielke, M., Bruck, R.: Sampling process information from unstructured data. In: Advanced Semiconductor Manufacturing Conference (ASMC), 2011 22nd Annual IEEE/SEMI, pp. 1–6. IEEExplore, New York (2011). doi:10.1109/ASMC.2011.5898184
17. Senturia, S.D.: Microsystem Design. Kluwer, Boston (2001)
18. Wagener, A., Popp, J., Schmidt, T., Hahn, K., Brück, R., Ortloff, D.: Environment for design and verification of mems fabrication processes. In: Proceedings of the MST 2005. VDE Verlag GmbH, Berlin (2005)
19. Yole Développement: Status of the MEMS Industry. Yole Développement, Lyon-Villeurbanne (2010)

Chapter 2
Introduction to Product Engineering

Abstract This chapter gives a general introduction into the authors' understanding of Product Engineering in general and defines the essentials of new product development and innovation. Several established baseline methods for product development and project management are presented and evaluated for their applicability. Therefore this chapter provides the base knowledge needed for the MEMS Product Engineering methodology presented in Part II of this book.

2.1 What is Product Engineering?

Before discussing the specific role of Product Engineering in the MNT industry and why customer orientation and distribution are predominant factors, it is necessary to clearly define what exactly comprises Product Engineering. In [20] Product Engineering is defined as "the discipline of driving and managing product development efforts following a predefined methodology" and that "it considers the whole realization-cycle of a product from its idea over the various development stages to a marketable product." Besides this generic definition there are more, slightly different, definitions for various disciplines (manufacturing, software development, etc.) reported in literature [5, 22]. All of these definitions have in common that they consider Product Engineering to be the task that drives product innovation from the product idea through the whole product life cycle (design, development, manufacturing, usage, and recycling or disposal).

To understand the boundary constraints of new products and to get a clearer picture, it needs to be defined what characterizes product innovation. Drivers for innovation and for new products are technology advances, changing customer needs, etc. Another area to be investigated are the key ingredients for successful new product developments and innovation in general. As documented by Cooper et al. in [6], their analysis revealed that there are several key ingredients:

- to embrace a long-term commitment to new product innovation,

D. Ortloff et al., *MEMS Product Engineering*, DOI 10.1007/978-3-7091-0706-5_2,

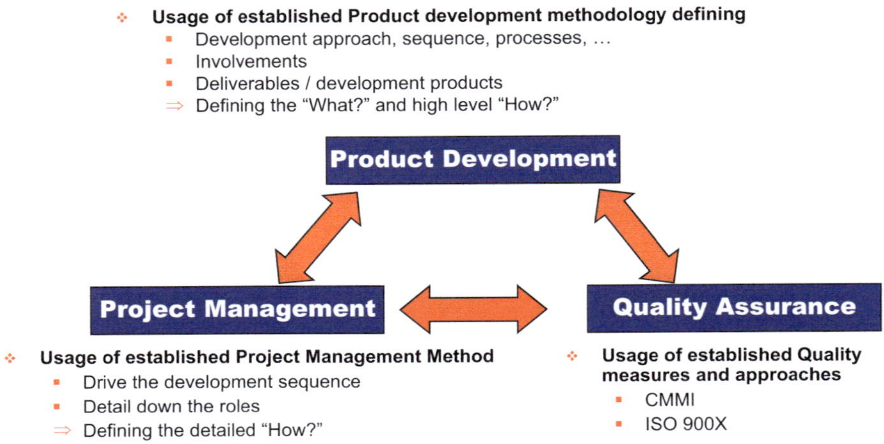

Fig. 2.1 Interactions between product development, project management, and quality assurance

- to develop a vision, objectives, and strategy for new product development efforts driven by and linked to the business's corporate objectives and strategy,
- to install a systematic, high quality new product process.

Several of the success factors are non-systematic and depend on company politics and management decisions. This book focuses mostly on the last point to establish a systematic new product process. In this book such an approach is referred to as a Product Engineering methodology.

The next question to answer is about the necessary ingredients for a high quality product engineering process. The first major ingredient is to have a methodology to coordinate the life cycle of a product. This mostly consists of processes for the development, the improvements, and the phase-out of a product. Because a pure methodology only sets the processes how to perform the tasks, the second necessary ingredient consists of a framework of tools supporting the different activities. Both, the methodology as well as the tool framework, require a laser sharp on quality. This means the quality of business process execution and the metrics/KPIs (Key Performance Indicators) thereof as well as a tool framework assuring the quality execution of the work and the product. All these aspects require high attention to develop high quality products. This triangle of forces and ingredients is depicted in Fig. 2.1.

The major challenge to be addressed for setting up product engineering activities is to select or define a product development methodology. This is a key activity and depends on the market, the business model, and the history of the company.

2.2 Product Engineering Essentials

To define a product development methodology and to set up the necessary method framework, it is essential to recap the essentials for product development / innovation. The following sections are only very brief summaries and much more detail can be found in, e.g., [6, 26].

2.2.1 Drivers for Innovation

Innovation is driven by several different factors which need to be reflected in the product creation process. These factors can be roughly categorized into technology advances, changing customer needs, shortening product life cycles, and increased world competition. Additionally a new product can be new to the company developing it or even new to the market. In detail new products can be split into the categories of new to the world products, new product lines, additions to existing product lines, improvements and revisions to existing products, repositionings (products developed for one area but afterwards used in a different area, e.g., Aspirin, Viagra) and new products due to cost reductions. Because these categories have a major impact on the business case of a potential new product, an idea for a product needs to be carefully analyzed and categorized during the initial assessments.

An additional aspect of innovation is that the best inventor is not necessarily the one who solves a problem best, it is often about seeing and understanding a problem. The detection and understanding of the customers' problems is key for an innovative company. Therefore the early involvement of customers in product ideas and a constant contact with the customers to detect problems is the key for future innovation. These points are supported by Davidsen et al. [7].

2.2.2 Success Factors for New Product Developments

Cooper and his colleagues have analyzed several different product development efforts and approaches. They report, in the widely accepted and supported book "Product Leadership" [6], the following key ingredients for new product success:

- Embracement of a long-term commitment to product development.
- Development of a vision, objectives, and strategy for the new product effort driven by and linked to the business's corporate objectives and strategy.
- Installation of a systematic, high quality new product process in the business and to practice discipline, following the principles of the process.
- Making available necessary resources.
- Fostering the innovation in the organization.

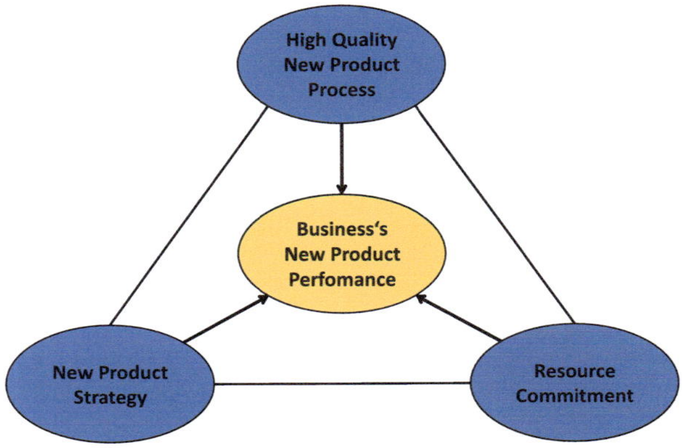

Fig. 2.2 Cornerstones of new product performance (based on [6])

These findings can be summarized into the cornerstones of new product performance shown in Fig. 2.2. It presents that the resource commitment, a new product strategy, and a new product process are the major contributors to a businesses innovation performance. The cornerstones business strategy and resource commitment are beyond the scope of this book and highly depend on the business itself. The third cornerstone of a systematic, high quality new product process is addressed by the Product Engineering Methodology described in Sect. 5.2.

2.2.3 *Success Factors for a New Product Process*

Before going into the details of the methodology, the key success factors for a new product process or product development methodology should be summarized as they are reported in literature. Cooper's book [6] presents several primary and secondary success factors for a new product development methodology. From their point of view a new product process needs to address the following goals:

- Exemplary Quality of Execution
- Sharper focus, better project prioritization
- Strong market orientation
- Better up-front homework and sharp, early, and stable product definition
- Fast-paced parallel processing by different disciplines
- A true cross-functional team approach
- Products with competitive advantage
- A fast-paced and flexible process

For an *Exemplary Quality of Execution* the parameters focusing on completeness, quality, and importance should be continuously watched. This can be achieved by establishing quality-control checkpoints, by designating the leadership team to become quality controllers, by clear and consistent metrics at intermediate checkpoints, by defining activities, tasks, methods, and best practices into the development phases, by specifying visible deliverables for each phase, along with action standards and by having effective resource allocation methods for each development phase.

Often the product engineering performance of a business suffers from too many projects with not enough resources. Therefore *sharper focus and better project prioritization* is essential for innovation to strive. The product engineering process needs to allow to have project funnels rather than project tunnels. So there need to be tough (Go/Hold/Recycle/Kill) decisions.

To achieve a *strong market orientation* of the innovation efforts, marketing tasks need to be designed into the product engineering process. This includes to perform preliminary market assessments. Market research is to determine user needs and wants, competitive analysis, value-in-use analysis, concept testing, customer reaction and feedback during development, user test and field trials, test market or trial sell and market launch based on a solid marketing plan. Additionally the project executives and leadership team needs to ensure the execution of marketing tasks and provides the necessary marketing resource.

For commercially successful innovation *high quality up-front homework and sharp, early and stable product definitions* are essential. The up-front homework needs to define the product and its business case in a stable manner. It lays the baseline for a successful or non-successful innovation. Therefore the new product process needs to ensure the proper execution of the early development phases and stages.

Although the Stage-Gate® process by its nature is a sequential process, the practical implementation of the product engineering process needs to enable and foster parallel execution of all necessary, multifunctional tasks. *A fast-paced parallel processing by different disciplines* is essential. Therefore a sequential approach has to be discouraged and each result achieved by one group needs to be passed on to the other groups as soon as possible, although the work might not be finished yet. Exchange of intermediate results and close cooperation is essential and should be fostered by the method.

Another essential feature of the methodology is *a true cross-functional team approach*. Such an approach needs the ingredients: A cross-functional team with committed team players; defined team captain or leader, accountable for the entire project; a leader with formal authority (co-opting authority from the functional heads); fluid team structure, with new members added or dropped as work demands; a small core group responsible, committed, and accountable from beginning to end.

To develop *Products with a competitive advantage / differentiated products* the development results need to deliver unique benefits and a superior value for the customer. To achieve this in an innovation process the required ingredients are: checkpoint criteria focusing on product superiority; development phases need activities fostering superiority; teams are required to deliver superiority evidence to checkpoints.

The product engineering process needs to enable fast progression and needs to accommodate for different types of projects. Therefore *a fast-paced and flexible process* is required. This implies that smaller or less risky projects need to be able to reduce the overhead by skipping or lightening certain managerial aspects but still execute the same process. The process must allow for flexibility within certain ranges and ensure fluidity in the process execution. For that fuzzy checkpoints need to be used, where decisions to proceed are taken although not all necessary deliverables are completely available but will be within a certain time frame.

The evaluated success factors inducing the above goals and additional factors for the business environment itself supporting the development methodology are given in [6].

2.3 Methods and Tools

Literature describes a plenitude of different product development methods, often targeted specifically at a certain industry like the Agile Method, the V-Model, and the Extreme Programming (XP), which are targeted specifically at the software industry. A recent and good summary of different approaches especially for user-driven product design is given in [14]. An older article summarizing several methods is [15].

Common ground of all those different approaches is the move to multidisciplinary teams and taking the whole life cycle of the product into account. Additionally, most modern methods can be categorized into either linear staged methods based on activities or into iterative, information and decision-based methods. The linear staged methods have a relatively fixed set of Stages. The iterative methods cycle the whole development loop several times. A good example for the linear staged methods is the Stage-Gate® method introduced in [6]. The Integrated Product and Process Development (IPPD) method, introduced in [8], is an example for an iterative method.

A background article on product development methods that compares Leonardo DaVincis problem detection and solving approach with current product development approaches can be found in [16]. The modern method used for the comparison is similar to the Stage-Gate® method and fosters multidisciplinary team structures. An additional similarity is the multiplicity of available tools at hand to the development team(s). The Leonardo DaVinci approach as well as the modern development approaches like Stage-Gate® consider the problem solving or product development as a combination of art and engineering. Both product development approaches try to foster the creativity through a multiplicity of tools available at hand.

2.3.1 Baseline Product Development Methods

2.3.1.1 IPPD/CE

The Integrated Product and Process Development (IPPD) approach is an expansion of the concurrent engineering (CE) approach. Both utilize a systematic approach to the integrated, concurrent development of a product and its associated manufacturing and sustainment processes. Their major aim is to satisfy the customer needs.

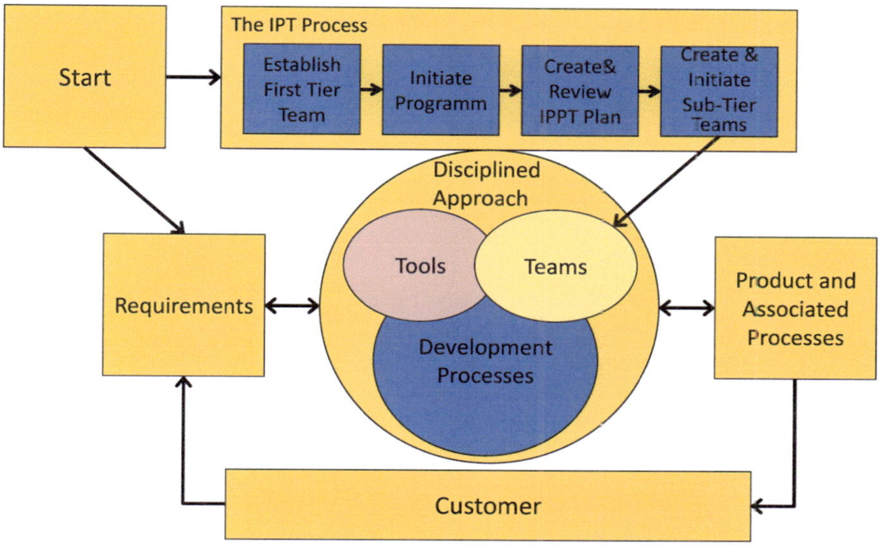

Fig. 2.3 A generic IPPD iterative process (based on [8])

IPPD is defined as: "A management process that integrates all activities from product concept through production and field support, using a multifunctional team, to simultaneously optimize the product and its manufacturing and sustainment processes to meet cost and performance objectives." [8]

A graphical overview of the generic IPPD management process is presented in Fig. 2.3. The input to the iterative development process shown in the figure is the requirements generated by the customer(s). They are elicited in a negotiation among many parties, each having serious and important concerns and impacts on the product. For that the customer(s) and their requirements need to be known in detail. The disciplined approach includes five general activities:

- understanding the requirements,
- outlining the approach,
- planning the effort,
- allocating resource and
- executing and tracking the plan.

Decisions made using this approach should be reevaluated as a system matures and circumstances (budgetary, threat, technology) change. The disciplined approach provides a framework for utilizing Tools, Teams, and Development Processes in a structured manner. Tools include documents, information systems, methods, and technologies that can be fit into a generic, shared framework that focuses on planning, executing, and tracking. Teams are made up of everyone who has a stake in the outcome or product of the team, including the customer and suppliers. Collectively, team members should represent the know-how needed and have the

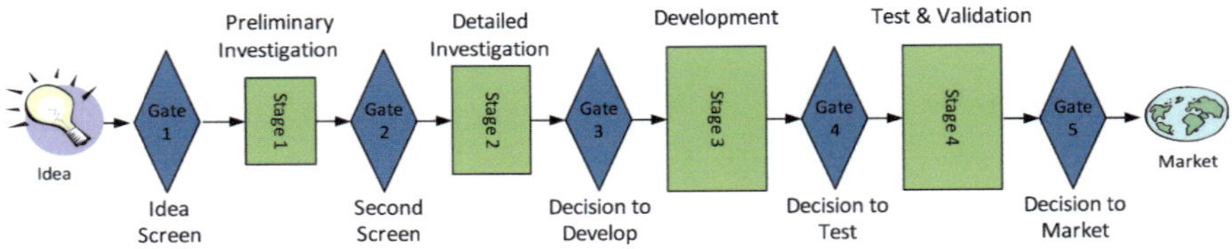

Fig. 2.4 The generic Stage-Gate® Process (based on [6])

ability to control the resources necessary for getting the job done. Development processes are those activities which lead to both, the end product and its associated processes. To ensure efficient use of resources, it is necessary to understand what activities are essential and how they affect the product and each other. Examples include requirements analysis, configuration management, and detailed design drawings. Product and Associated Processes include what is produced and provided to the customer. These steps are iterated if additional or changed customer requirements are encountered.

In summary IPPD and CE are development processes highly focused on the customer requirements and fulfilling them by iterative product extensions and releases. In so far the basic approach is rather similar to the Agile development method for software. An overview of the approach and its application in the acquisition process of the Department of Defence is given in [8].

2.3.1.2 Stage-Gate®

The Stage-Gate® process is an operational roadmap for driving new product development projects from idea to launch—a blueprint for managing the new product process improving effectiveness and efficiency. The central point of this method is a staging and gating process where each Stage is executed by a multidisciplinary team and the Gates build intermediate assessment points. Pioneered and developed by Robert G. Cooper, published in the book [6] and several papers and web sites, it is the world's most widely used new product development process. Examples for adopters of the Stage-Gate® methodology are Hewlett-Packard, Procter&Gamble, and SonyEricsson. An overview of the generic process is presented in Fig. 2.4.

A successful innovation begins with an idea and ends with the launch of a new product. The steps between these points can be viewed as a dynamic process. Stage-Gate® divides this process into a series of activities (Stages) and decision points (Gates) covering the success factors described in Sect. 2.1.

Each Stage contains a set of defined concurrent activities, incorporating industry best practices. Activities during each Stage are executed in parallel to enhance speed to market. This is further enhanced by the use of cross-functional/multidisciplinary teams all working towards the same goals and the same set of deliverables required for the next Gate. As the method is a blueprint for new product processes, the amount of Stages can vary between different process implementations in different environments. Typical process implementations use growing sizes of Stages where each later Stage is bigger than a previous one in terms of commitment, effort, money, etc. On the other hand the risks and the uncertainties of the project need to decrease.

Gates are the decision points where the Gatekeepers decide whether to continue funding a project. The Gates consist of three main elements which are the required deliverables as input to the Gate, a set of evaluation criteria for the deliverables and the outputs. The outputs, in the positive assessment case, consist of the funding decision, approved deliverables, an agreed action plan for the next Stage and an agreement on the set of required deliverables for the next Gate. At the Gates selection criteria are applied and tough decisions are made about the projects, but the Gates are essential for the success of a new project strategy. They are the keys to distinguish between the potentially good and bad products and enable to focus scarce resources on the potentially best projects, thus establishing value for money. The outcome of the decision point is either to Go (continue developments), Kill (cancel developments), Hold (temporarily stop developments), or Recycle (redo previous Stage) the project. Gates also act as "quality control" checkpoints evaluating among other things: Have the deliverables been generated in a quality fashion? Is the project (still) attractive from an economic and business standpoint? Is the information sufficient to make a decision? Is the action plan and request for resources sound?

The Gates are executed by the Gatekeepers of the process. The Gatekeepers are a team of senior leaders who make Go/Kill/Hold/Recycle decisions at the Gates. They are from different functional areas and can commit the required resources. They use a pre-set list of criteria and rules for the Gates. By those rules and criteria they are bound so that they cannot play favorites.

The benefits of using Stage-Gate® are that it introduces discipline into an ordinarily chaotic process. It focuses attention upon quality of execution and speeds up the innovation process. The method ensures a complete process (no critical steps are omitted) and facilitates a focus on the performance of a new product. Software environments supporting and guiding developments based on this method are available as well, one example is PD-Trak [9].

2.3.1.3 Selection

As regularly reported in literature, there is nothing like "The method" fitting all purposes and industries. Therefore the application area of the product engineering approach / product development method needs to be carefully taken into account. The method developed in the scope of the CORONA project and this book is to be used in the MNT area. Advantages of the one or the other method may be less relevant in the considered case. As an example, certain aspects of the criticism about the staged development given in [2] about the missing integration of all life cycle aspects are less relevant in the MNT space due to the nature of the value chain and that MNT products are only a small portion of a final, tangible customer product.

Aspects claimed as advantages of the IPPD methodology like the inclusion of manufacturing aspects, distribution and inclusion of suppliers into the engineering process can easily be covered with the extensions through the T^2M method as well as through the multidisciplinary teams approach in the MNT field.

Another specialty to take into account is the sequential nature of the MNT process developments (introduced in more detail in Sect. 4.3.1). During the process development efforts the steps material development, material optimization, sequencing, modularization, etc. need to be executed in this standard sequence and can hardly be changed or parallelized. Therefore the process development has a kind of predefined nature and needs to be incorporated. These efforts can be integrated more easily into the Stage-Gate® method.

Furthermore the CORONA project executed a market analysis about the currently used methodologies in the MNT field. An overwhelming majority of currently in use methods, where a formal method was in place, were based on a phase and gate-based process anyway. The independent efforts of the MEMS Industry Group (MIG) to develop the MIG Technology Development Process (TDP) template [17] resulted in exactly the same results. As highlighted in the whitepaper describing the template, their survey on common industry practices concluded that the most widely used development methodology is based on a staged and gated-based process. Actually the process proposed by the template is rather similar to the process suggested by this book.

MNT product innovation has a very high research character. Due to that the "contract" between the management and the engineering team, being the essential base in IPPD, is very difficult to achieve because of all the unknowns. Additionally, considering the engineering with different teams along the value chain, the methodology needs to cater for structure, possible management and partner involvement, team building and multi-site / multi-partner development. Therefore a highly integrated development method like IPPD requiring a co-located development team and offering only less management involvement possibilities has several disadvantages. Especially the required multi-site / multi-partner engineering team of the MNT business, the technology development needs to be done at least at one place with equipment access, requires a more structured and controlled approach.

Therefore it can be concluded that the advantages of IPPD in comparison with Stage-Gate®, as it has been discussed in [2], are less applicable. Additionally the arguments for IPPD have some drawbacks as well, especially to the MNT market. As the author suggests in the showcase, the transition from a Stage-Gate® method to an IPPD-based method is possible as well and to structure the sometimes chaotic process, a staged method is preferable. Additionally Stage-Gate® has a very good reputation, is in widespread use, and is supported by various additional authors [7, 19]. Because of that the decision was taken to base the herein proposed methodology on the Stage-Gate® approach, having the preconditions for IPPD in mind and working around some issues via the T^2M extension.

2.3.2 *Baseline Project Management Methods*

The necessity for the addition of a Project Management (PM) method extending the Product Development method is described in the methods overview (Sect. 2.1). To not digress too much from the subject, no general introduction into project management itself is given herein. For such an introduction and the historical progression of project management, e.g., [23] can be reviewed. Especially [23] gives a good definition for what a project is, which success factors do exist, and what distinguishes a project from a product, program, and operations.

A literature study revealed that there exist a variety of different PM Methods. An exhaustive list of such methods and project maturity models can be found in [12]. An analysis of how commonly the different methods are used and their general applicability to product developments of all kinds, not limiting it to software projects, reduces the possible candidates. Additionally it must be easily possible to combine the PM method with a product development method. Applying all these criteria the remaining list consists of PMBOK [11], PRINCE2® [18], and ISO10006 [13].

To give an overview, the following sections very briefly summarize these three methods and the differences between them.

2.3.2.1 PMBOK

The Project Management Body Of Knowledge (PMBOK) is standardized in the IEEE Std. 1490–1998 and can be considered *the* project management standard in North America. It is officially documented in [11].

Under PMBOK a project is initiated, guided, etc. by its Stakeholders. These are the people involved or affected by the project activities, e.g., project sponsor, project manager, user, suppliers, and customers. They define the needs, provide resources and so on. The PMBOK project management principles use five main PM processes which are initiating, planning, executing, controlling, and closing of the project. The processes partly overlap each other as well as they interact with each other. An overview of the interaction of the processes is presented in Fig. 2.5. PMBOK describes the processes in terms of Inputs (e.g., documents, plans, designs), tools & techniques (mechanisms applied to the inputs), and outputs (e.g., documentations, products).

Additionally the standard defines nine knowledge areas which are closely tied into the processes and the tools, techniques, and best practices. The nine knowledge areas are heading the different rows in Table 2.1 and the columns indicate the different process groups. The cross sections of the table list the corresponding processes of the group in that knowledge area and refer to the corresponding PMBOK chapter. A graphical overview of the interaction of the different knowledge areas is presented in Fig. 2.6.

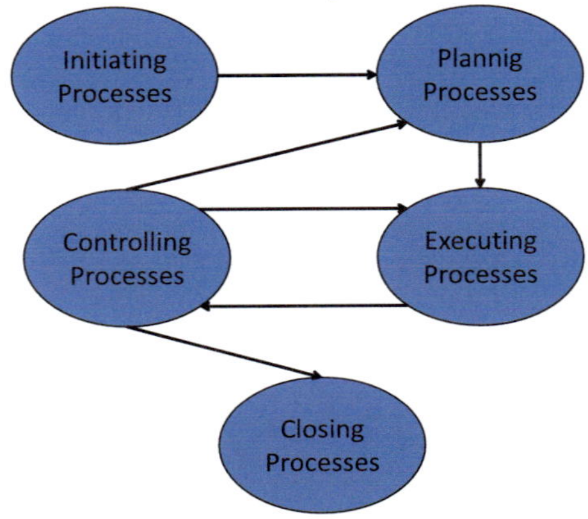

Fig. 2.5 Management process of the PMBOK (based on [11])

Table 2.1 PMBOK overview

Knowledge Areas & Chapter numbers	Process Groups & Processes Within Groups Process section no. within PMBOK Chapter				
	Initiation	**Planning**	**Executing**	**Monitoring & Controlling**	**Closing**
Project Management Integration 4	1 Develop Project Charter 2 Develop Preliminary Project Scope Statement	3 Develop Project Management Plan	4 Direct and Manage Project Execution	5 Monitor and Control Project Work 6 Integrated Change Control	7 Close Project
Project Scope Management 5		1 Scope Planning 2 Scope Definition 3 Create Work Breakdown Structure (WBS)		4 Scope Verification 5 Scope Control	
Project Time Management 6		1 Activity Definition 2 Activity Sequencing 3 Activity Resource Estimating 4 Activity Duration Estimating 5 Schedule Development		6 Schedule Control	
Project Cost Management 7		1 Cost Estimating 2 Cost Budgeting		3 Cost Control	
Project Quality Management 8		1 Quality Planning	2 Perform Quality Assurance	3 Perform Quality Control	
Project Human Resource Management 9		1 Human Resource Planning	2 Acquire Project Team 3 Develop Project Team	4 Manage Project Team	
Project Communications Management 10		1 Communications Planning	2 Information Distribution	3 Performance Reporting 4 Manage Stakeholders	
Project Risk Management 11		1 Risk Management Planning 2 Risk Identification 3 Qualitative Risk Analysis 4 Quantitative Risk Analysis 5 Risk Response Planning		6 Risk Monitoring and Control	
Project Procurement Management 12		1 Plan Purchases and Acquisitions 2 Plan Contracting	3 Request Seller Responses 4 Select Sellers	5 Contract Administration	6 Contract Closure

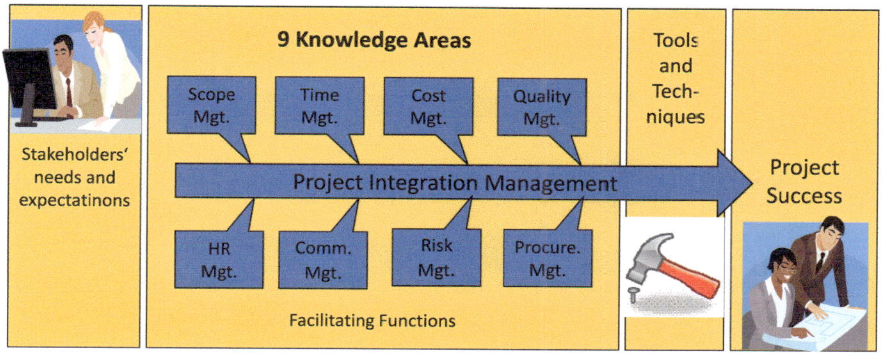

Fig. 2.6 Knowledge areas and their interaction of the PMBOK (extracted from [23])

In summary PMBOK can be characterized to be a collection of processes and knowledge areas generally accepted as best practices within PM. Its basic concepts are applicable to all projects and programs (and operations). The knowledge base provides the fundamentals of PM, irrespective of the type of project (software, construction, environmental, aerospace, etc.) and recognizes five basic process groups and nine knowledge areas typical for almost all projects.

2.3.2.2 PRINCE2®

The PRINCE2® project management method, published and maintained by the Office of Government Commerce (OGC) in the UK, is widely used and has become the standard for project management in Europe. The OGC has published several books on PRINCE2® with [18] being the most important one. Another good resource about the method is [3].

PRINCE2® is based on the approach to treat the efforts of project management as a process. It is highly product oriented by focusing on the business case and emphasizing quality aspects. The method provides a framework of components and techniques to perform the different project management processes. The controlled and systematic preparations and execution of projects in a process following PRINCE2® explicitly prevent problems of other PM methods with weaker processes. This is a central aspect and an additional benefit compared to other methods. An overview of the PRINCE2® processes, its components, and the techniques provided is presented in Fig. 2.7. The figure shows the interaction between the different processes depicted in Table 2.2. In summary PRINCE2® can be characterized to be a collection of processes, components, and techniques generally applicable to PM in diverse fields (software, construction, environmental, aerospace, etc.). Its basic concepts are applicable to all projects and programs (and operations). The method provides the fundamentals of PM, recognizing eight basic processes with sub-elements, eight components used during the different processes, and three basic techniques applied.

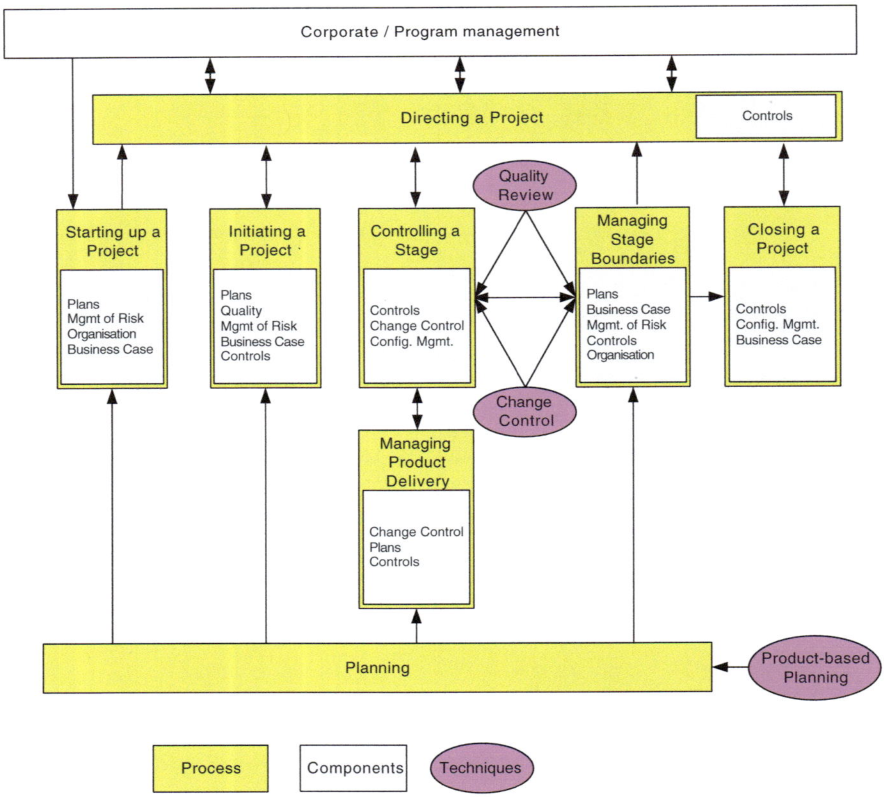

Fig. 2.7 Management processes, components, and techniques of PRINCE2® (based on [3])

The PRINCE2® method benefits can be summarized as, the method

1. identifies management, specialist, and quality Products/Deliverables and helps to ensure that they are produced on time and to budget
2. focuses attention on the quality of Products/Deliverables and on the viability of the business case
3. separates the management and specialist aspects of organization, planning, and control
4. facilitates controls at all levels
5. makes the project's progress more visible to management
6. provides a communication medium for all project staff
7. ensures that work progresses in the correct sequence
8. involves senior management in the project at the right time and in the right place
9. allows the project to be stopped and, if required, re-started completely under management control, at any time in the project's life
10. is in the public domain and requires no license fee
11. has a well-established user group dedicated to the support, promotion, and strengthening of the method

Table 2.2 PRINCE2® overview

Process	Process elements	Components	Techniques
Starting up a project	SU1. Appointing a project executive and a project manager SU2. Designing a project management team SU3. Appointing a project management team SU4. Preparing a project brief SU5. Defining project approach SU6. Planning initiation stage	Plans Mgmt. of risk Organization Business case	
Directing a project	DP1 Authorizing initiation DP2 Authorizing a project DP3 Authorizing a stage or exception plan DP4 Giving ad hoc direction DP5 Confirming project closure		
Planning	PL1. Designing a plan PL2. Defining and analyzing products PL3. Identifying activities and dependencies PL4. Estimating PL5. Scheduling PL6. Analyzing risks PL7. Completing a plan	Product-based planning	
Initiating a project	IP1. Planning quality IP2. Planning a project IP3. Refining the business case and risks IP4. Setting up project controls IP5. Setting up project files IP6. Assembling a project initiation document (PID)	Plans Quality Mgmt. of risk Business case Controls	
Controlling a stage	CS1. Authorizing work package CS2. Assessing progress CS3. Capturing project issues CS4. Examining project issues CS5. Reviewing stage status CS6. Reporting highlights CS7. Taking corrective action CS8. Escalating project issues CS9. Receiving completed work package	Controls Change control Config. Mgmt.	Quality review Change control
Managing product delivery	MP1. Accepting a work package MP2. Executing a work package MP3. Delivering a work package	Change control Plans Controls	Quality review Change control
Managing stage boundaries	SB1. Planning a stage SB2. Updating a project plan SB3. Updating a project business case SB4. Updating the risk log SB5. Reporting stage end SB6. Producing an exception plan	Plans Business case Mgmt. of risk Controls Organization	
Closing a project	CP1. Decommissioning a project CP2. Identifying follow-on actions CP3. Project evaluation review	Controls Config. Mgmt. Business case	

2.3.2.3 ISO10006

Another standard in the context of project management is ISO10006 which is documented in [13]. Up to now the standard has been released in two revisions, ISO10006:1997 and ISO10006:2003 [4]. The initial version of this standard is based on a draft of PMBOK and is therefore relatively similar to PMBOK. But the ISO standard addresses project management from the quality aspects and targets more to quality assurance in project management. On top of that it is less detailed than PMBOK and offers less help to project managers. Being based on PMBOK it shares the weakness of being more a collection of best practices, knowledge areas, and required skills. The ISO10006 focuses less on an explicit project management process and has only a very abstract and loose project execution process. Therefore it is less practical and provides more a kind of a checklist for quality-driven project management than a project management process. An additional weak point of the ISO10006 is that it misses or only very briefly covers subjects like scope, cost, time, and risk. Literature comparing PMBOK and ISO10006 [25] or analyzing ISO10006 itself [21] discourages the use of ISO10006. The later article even states that the application of ISO10006 could result in the opposite effects because of potential misinterpretations of some guidelines having no clear context. Because of these weaknesses and the very limited usage of ISO10006 in commercial environments this method is disregarded from further investigation.

2.3.2.4 Selection

Only PMBOK and PRINCE2® remain for consideration as a baseline project management method, after having eliminated the ISO10006 standard because of its weaknesses and its focus on quality management in projects. Investigating these two possibilities further it can be noted that PMBOK is less process centric, although it defines the five basic processes. PMBOK puts a lot of emphasis on the different knowledge areas a project manager needs to handle and approaches project management from the project manager's required skill set. This approach makes the PMBOK method very valuable for teaching project management skills. In contrast PRINCE2® is focused more on the practical guidance of a particular project. This point of view is shared by [27, 28]. This process and guidance orientation, together with the inherent management of the project in stages, makes PRINCE2® a better fit with the Stage-Gate® method than PMBOK.

An additional difference between PMBOK and PRINCE2® is that the latter emphasizes the business case and the project brief and PMBOK does not even mention one of them. The business case is a very central vehicle for the PRINCE2® method. The method enforces the constant update according to gained knowledge and changing conditions as well as its constant validation by the project board after each development stage. Therefore PRINCE2® has a higher applicability for product engineering than PMBOK. This is even emphasized further by the product, rather than task-centric planning method of PRINCE2®. A product in that context can be any kind of outcome of a project, a tangible one, like an MNT device, or an intangible one like a redefinition of a business process. On the other hand PMBOK contains knowledge areas like human resources (HR) and procurement management which are not covered in PRINCE2®.

Another distinguishing aspect is that PRINCE2® is designed for a variety of customer and supplier scenarios while PMBOK primarily considers that the project is run by one supplier for one customer. This aspect makes the PRINCE2® method more flexibly applicable, especially for scenarios where the project is run by an instance in a sandwich position having customers and external suppliers.

Furthermore it can be noticed that PRINCE2® puts more emphasis on documentation than PMBOK. It provides a plentitude of templates or guidelines for project templates helping to become the knowledge pool for a project. Between those are the definitions of an issue log, a risk log, and acceptance criteria that are not addressed by PMBOK. On the other hand this strong emphasis on documentation can burden a project and creates a significant management overhead for small projects. This can be prevented by using PRINCE2® as a framework where a project can select the applicable standards depending on risk and size.

Summing up and concluding, it can be said that PRINCE2® is a better extending match for the Stage-Gate® method than PMBOK. This is due to its advantages in its process model, its product and business case-centric definition, and its more flexible project environment scenario. Its drawbacks of missing out certain areas, which are covered by PMBOK, can be circumvented by adding not covered ground from the PMBOK and using its more detailed documentation of best practices. This combined approach is suggested by [24, 27] as well. The documentation overhead can be reduced by relaxing the constraints for small and minor risk projects. On the other hand the focus on documentation builds a good baseline for quality certification.

2.3.3 *T²M*

As noted in the last section an increased involvement of the customer in the Product Engineering process is necessary. On the other hand there is also an increased need for modularity and flexibility. Every company has developed its own business processes and cannot be forced to fully adapt to external processes introduced by random customers. The goal is therefore to devise a highly customizable methodology for customer interaction that doesn't disrupt or damage established processes.

The overall approach of CORONA on Product Engineering is deliverable driven instead of task driven: "you can't control what you can't manage, and you can't manage what you can't specify" [1]. Meaning that the progress of the project is measured based on concrete artefacts (deliverables) created instead of workload/money spent. For these reasons the CORONA approach supports a more fine-granular handling of customer interaction and collaborative engineering. The Task Transition

Methodology (T^2M) provides the means for organizing the engineering work within the Stages. T^2M distinguishes between three basic entities:

Tasks—describing product creation activities which are implementing
Deliverables—which are documents or other artefacts that are checked at
Transitions—serving as synchronization points between tasks

In short Tasks generate Deliverables that are checked at Transitions that again lead to subsequent Tasks. At first glance T^2M looks quite similar to Stage-Gate®. It is definitely not the same as it will become clear when defining the concrete semantics of Tasks, Deliverables, and Transitions.

2.3.3.1 Tasks

In T^2M Tasks represent activities that implement Deliverables or to put it in less abstract words: Tasks are activities whose results are called Deliverables, e.g., the Deliverable of a Task "Market Analysis" would be a "Market Analysis Report." Therefore the most basic Task consists of a name that describes the overall activity, a list of Deliverables that have to be implemented, and last but not least the budget that has been assigned to perform the Task.

Additionally, T^2M Tasks may serve as container entities that contain other (Sub-) Tasks and Transitions. This allows building up a hierarchy, which could, for example, be used to define methodology templates. It is not intended to use the hierarchy for micro managing. Upcoming issues within a Task are not escalated outside of the Task. To the outside every Task is strictly a black box.

The size and scope of each Task depends on the preferences of the involved partners. Very small Tasks lead to larger overhead and limited flexibility and should generally be avoided in projects where innovation is important. Too large Tasks are harder to control and may commence in unwanted directions. As a rule of thumb Tasks should be split at significant milestones or if the executing personnel changes.

While it is in principle possible that more than one company is involved in a single Task it is unadvisable with regard to two of the main challenges in MNT Product Engineering: different locations and IPR. It is therefore advisable to split a Task into sub-tasks till every sub-task can be performed at a single location/company and the same personnel involved.

2.3.3.2 Deliverables

The Deliverable of T^2M is the actual outcome of a Task. Like in Stage-Gate® this has to be something tangible like a document, a computer program, or a physical artefact. Spent money or wasted time may be the main outcome of some Tasks but do NOT count as Deliverables. An additional restriction of T^2M is that it forbids Deliverables that create overhead. There may never be a Deliverable that is created

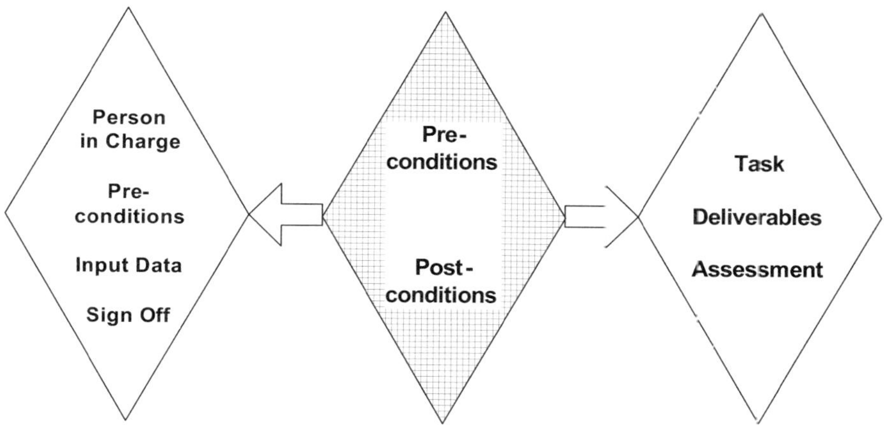

Fig. 2.8 Transition with pre- and post-conditions

only for the sake of a Transition. A Deliverable should be created to advance the project and not the process.

Each Deliverable has a specification, an implementation, and a rating. The specification describes the requirements and constraints of the Deliverable. The implementation is the resulting Deliverable, and the rating describes how the implementation matches the specification. In its most simple form this could be "accepted" or "rejected."

A Deliverable is always implemented during a single Task. While under practical conditions it is possible to share an incomplete Deliverable with other Tasks the responsibility for completing the Deliverable stays at the original Task.

2.3.3.3 Transitions

The Transition serves as a synchronization point between Tasks. Here the results (Deliverables) of the preceding Task are collected and handed over to the subsequent Tasks. A Transition is activated if all required Deliverables are accepted. After activation of a Transition all preceding Tasks are closed and all subsequent Tasks are activated. This procedure and their pre- and post-conditions are presented in Fig. 2.8. Like Janus, the Roman god of Gates, each transition has two faces. One that looks back at the previous Tasks and checks if all specified Deliverables are accepted and another one that looks at the upcoming Task to specify the next Deliverables.

The Transition is controlled by the persons in charge of the subsequent Task(s) and, depending on individual agreements, the customer. This is reasonable as those persons are also the ones that ought to use the Deliverables of the last Task and have sufficient knowledge of the Task to provide an implementable specification.

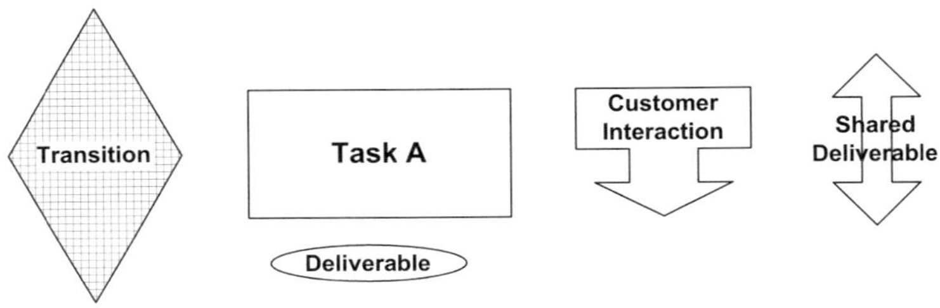

Fig. 2.9 5 Symbols used in T^2M diagrams

It should be noted, however, that the actual specification of the Deliverables does not take place during project execution but during project planning. The Transition provides only an opportunity for last minute changes to the specification! A Transition is also the only place where customer involvement is allowed (if the customer is not actively involved in the Task anyway). Typical contributions of the customer are result assessments and specification updates.

As mentioned earlier, Transitions are similar but not equivalent to the Gates of Stage-Gate®. In contrast to Gates, Transitions are meant to be lightweight synchronization points between Tasks with possible customer feedback. There is no detailed assessment of the Deliverables, no hard decisions, and no budget approvals. Ideally there is not even a personal meeting. Should any issues arise the transition stays inactive till the issues are resolved by the PRINCE2® processes.

2.3.3.4 Modeling Processes with T^2M

T^2M processes are modeled by concatenating Tasks and Transitions. A T^2M diagram consists mostly of the four basic elements shown in Fig. 2.9.

The symbol labeled "Customer Interaction" can be added to Transitions to indicate that at this transitions the customer is involved. The actual nature of this interaction should be reflected in the symbol's label (e.g., "notification," "sign-off," "review"). The double-sided arrow labeled "Shared Deliverable" is used when two parallel Tasks are so closely related that a synchronization via Transitions is too inconvenient. In this case unfinished Deliverables or other documents may be shared on mutual agreement.

A very basic T^2M process with three Tasks and three Transitions is shown in Fig. 2.10. Task 2a and 2b run in parallel and are synchronized by Trans2 and Trans3. The Tasks either are used as a container for Sub-Tasks (e.g., other T^2M modules) or are handled as a black box. Whatever happens inside a black box, the persons in charge for that Task has the full responsibility for that Task, including potential Sub-Tasks. The first Transition of a module is called Entry-Transition and ensures that all preconditions of the module are met. Like any other Transition it is controlled by the persons responsible for the immediately following Tasks.

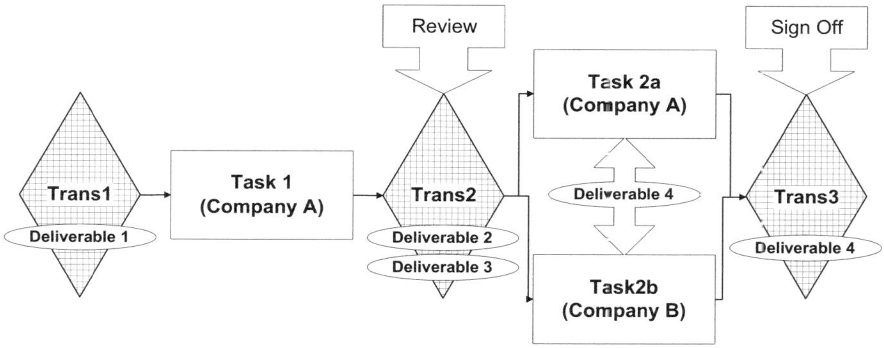

Fig. 2.10 Small example T^2M

The results (Deliverables) of a Task are collected in the Transition that directly follows the Task. Here the consistency of specification and implementation of each Deliverable are checked. Only if all Deliverables of a task are accepted the Transition is activated.

As noted in Sect. 2.3.3.3 a Transition is also the only place for customer interaction in a T^2M process. In a T^2M diagram customer interaction is modeled by placing a customer interaction symbol right above the Transition in question. The actual nature of the customer interaction depends on individual agreements and may range from simple status reports to full Transition control. In the example diagram shown in Fig. 2.10 this has been done for Trans2 and Trans3. The customer interaction in Trans2 would be a simple review of the deliverables, so the customer has the opportunity to assess the current status and to give input to the following Tasks. The customer interaction for Trans3 is labeled "sign-off" which essentially means that a sign-off of some or all of the Deliverables by the customer is required. As a result the Transition can only be activated if the customer allows it.

Besides process control and customer interaction the main objective of the Transitions in a T^2M process is Task synchronization. The synchronization aspect happens if multiple Tasks start or end at a single Transition. Some example scenario of these task synchronization aspects are presented in Fig. 2.11. More than one Task starting at a single Transition, as indicated in the upper left of Fig. 2.11, makes sense if the Tasks are closely related. Per definition a Transition is controlled by the persons responsible for the immediately following Tasks. Therefore there has to be a consensual agreement about the acceptance of the preceding Deliverables and the specification of the upcoming Deliverables. Otherwise the Transition wouldn't be activated and the Tasks wouldn't start. If there is a disagreement on the assessment which cannot be resolved between the people directly involved in the transition, the topic becomes escalated to the project management, according to the PRINCE2® processes.

If more than one Task ends at a single Transition, as indicated in the upper right of Fig. 2.11, the situation is similar. The Transition cannot be activated until all the

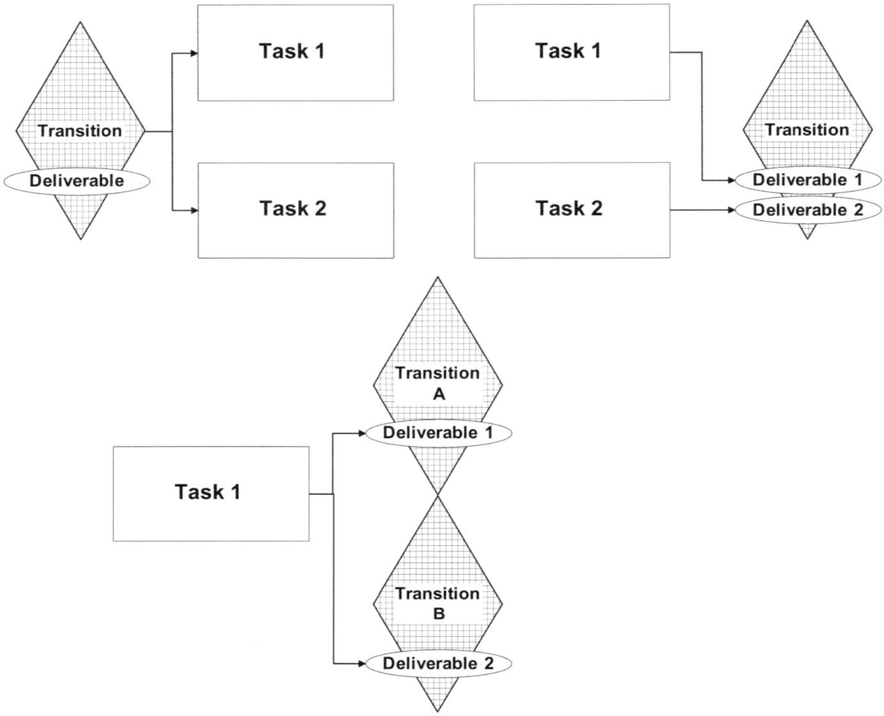

Fig. 2.11 T^2M task synchronization

Deliverables of *all* Tasks are accepted. As a result neither Task is closed, so work may continue till all interdependencies between the parallel Tasks have been sorted out. Otherwise one of the Tasks may end and the team disbanded, while the other Task still needs input.

Another possible situation is multiple Transitions following a single Task, as indicated in the bottom of Fig. 2.11. Such a situation becomes feasible if the Deliverables of that Task are used in completely independent follow-up Tasks. As a result some of the Tasks may already start while others are still waiting for their input Deliverables to be accepted. Of course the Task preceding the transition may only be closed when all Transitions have been activated.

Ideally every Deliverable is implemented during a single Task and only used by other Tasks after being accepted. In a real Product Engineering process such a clean and definite approach is often not feasible. Especially in parallel development tasks, e.g., MEMS- and IC-components, a need for very fine granular information exchange may arise. In a T^2M diagram such a tight coupling between two Tasks is represented by a "shared deliverable" symbol. A shared Deliverable means that while the responsibility for implementing the Deliverable still remains at the original Task, intermediate states of the Deliverable are shared with the other Tasks. As this

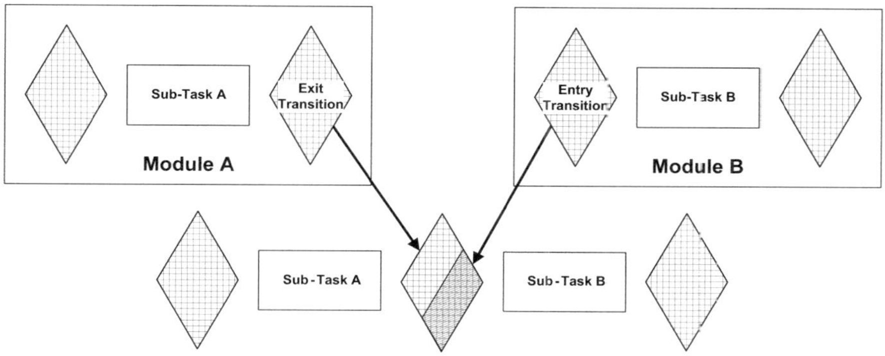

Fig. 2.12 Combining T²M modules

happens without T²M synchronization mechanisms, information consistency has to be provided by other technical or nontechnical means, e.g., by an electronic project binder.

2.3.3.5 T²M Modularization

This book aims to provide "Best Practices" for certain organizational situations that may occur during cooperative Product Engineering. To be able to model some of these practices using T²M processes a modularization concept is needed. Fortunately, modularization with T²M is quite straightforward following a few simple rules. A T²M module is basically a Task that contains at least two dedicated Transitions: The first Transition of a module is called Entry-Transition and ensures that all preconditions of the module are met. Like any other Transition it is controlled by the persons responsible for the immediately following Tasks. The main difference to "normal" Transitions is that there are no Deliverables to be accepted, which is quite logical as there is no preceding Task to implement them. The functionality of the Entry-Transition is therefore limited to the planning of the subsequent Tasks. The Exit-Transition, on the other hand, has no subsequent Task. Therefore the Exit-Transition lacks all aspects concerned with planning of subsequent Tasks. Modeling T²M processes from T²M modules therefore means combining the Exit- and Entry-Transitions of two neighboring modules into a full Transition as shown in Fig. 2.12.

Overall, because the T²M method is embedded into the Stage-Gate® method a quite hierarchical approach can be modeled in a consistent and comprehensive way. The Stage-Gate® method provides a general framework for innovation, guided by PRINCE2® project management process realized through breaking down the Stages into a product/project-specific T²M model for realization.

Table 2.3 Responsibilities in the preliminary investigation stage (example)

	R&D management	Development management	Manufacturing management	Design team	Process team	Package team	Processing engineers	Operators	M&S	Legal	Financial	Project management
Preliminary technical feasibility report	A	I		R	R	R	S					I
Preliminary market assessment report	C								R			I
Preliminary prospects report	C								R			I
Preliminary partner list	C								R			I
Preliminary legal assessment report	C			C	C	C				R		I
Draft project mandate	C			C	C	C						R
Project initialization stage plan	C			C	C	C						R
Req. for authorization to proceed												R

2.3.4 *Responsibility Assignment Matrix*

One of the most important tasks in modeling an innovation process is the assignment of responsibilities. Naturally in a strictly hierarchical work environment there is a kind of pyramid structure with a small number of high ranked executives at the top (CEO, CTO, etc.) and a larger number of project workers with a limited decision range at the bottom. For cooperative distributed engineering projects the situation is not that easy as the responsibility is shared among the various partner companies and the customer. The customer is in a special situation as, while on top of the hierarchy in our PE approach, he/she has basically no responsibility on the project results.

In our PE approach the responsibility is assigned and executed at decision points (Gates, Transitions) and connected with Deliverables. A common method to describe and identify responsibilities is the responsibility assignment matrix (RAM) or RACI-matrix, which describes the involvement of project partners in project Tasks and Deliverables. There are several ways to describe this involvement. Most common is the RACI method which classifies the involvement into one of the four categories *responsible (R)*, *accountable (A)*, *consulted (C)*, and *informed (I)*. It is described in more detail in [10, 23]. Responsible is used to describe the parties who are doing the work to achieve the Task. Accountable is the party who is ultimately accountable for the results of the work. There can be only one party accountable for any given Task or Deliverable. Consulted stands for parties who contribute with their experience or knowledge but are not directly involved. Informed stands for parties that are just informed about the current state or the results of the work. In contrast to consulted they are not expected to generate feedback. An example of an RACI matrix as used in Sect. 5.3.3 is given in Table 2.3.

There are several alternative approaches to RACI that extend the participation model and can be used to provide a more fine granular distinction of responsibilities. Most notably are RASCI (documented in, e.g., [10]), which adds a *support (S)* category that describes parties who are assigned to *responsible (R)* to assist in completing the work, and CAIRO, which adds a *omitted (O)* category for parties that are explicitly not to be involved.

Due to the complex nature of cooperative distributed engineering we distinguish between three hierarchical levels. The top-level describes how the responsibilities are shared among the project partners. The matrix documents the role each partner has at the project management board, the individual Gates and the top-level Deliverables (e.g., IC, MEMS, and package). This is clearly a level where no individuals but company departments appear on the matrix. The matrix distinguishes between the four partner types of customer, developer, supplier, and user. The middle level is mostly about individual executive responsibility. Here the contact-persons of each company/department are assigned to roles within project management, Gates and T^2M Transitions. The third level deals with the responsibilities of the individual developer that are linked to the implementation of concrete Deliverables.

References

1. Allen, P., Higgins, S., McRae, P., Schlamann, H.: Service Orientation: Winning Strategies and Best Practices, 1st edn. Cambridge University Press, Cambridge (2006)
2. Anderson, R.E.: Phased product development: friend or foe? Business Horizons **39**(6), 30–36 (1996)
3. Bentley, C.: PRINCE2 a Practical Handbook, 3rd edn. ButterworthHeinemann, Oxford (2009)
4. British Standard Institute: ISO 10006:2003: Quality management systems – Guidelines for quality management in projects (2003)
5. Buxton, W.: Performance by design: the role of design in software product development. In: Proceedings of the Second International Conference on Usage-Centered Design, pp. 1–15. Constantine & Lockwood Ltd., Rowley (2003)
6. Cooper, R.G.: Product Leadership – Creating and Lauching Superior New Products, 1st edn. Perseus Books, New York (1998)
7. Davidsen, B.A.: Innovation and Product Development - Methods and Tools. Telektronikk **100**(2) (2004)
8. Department of Defense: Dod Guide to Integrated Product and Process Development. Office of the Secretary of Defense (OSD), Washington (1996)
9. DRM Associates, PD-TRAK® NPD Process Tools, http://www.npd-solutions.com/pdtrak.html (2011). Accessed 24 Jan 2013
10. Hightower, R.: Internal Controls, Policies and Procedures, 1st edn. Wiley, Hoboken (2008)
11. Institute, P.M.: A Guide to the Project Management Body of Knowledge, 3rd edn. Project Management Institute, Newtown Square (2004)
12. Institute, P.M.: Organizational Project Management Maturity Model (OPM3), 2nd edn. Project Management Institute, Newtown Square (2013)
13. International Organization for Standardization: ISO 10006: Quality management systems – Guidelines for quality management in projects (2003)
14. Janhager, J.: User Consideration in Early Stages of Product Development - Theories and Methods. PhD Thesis, KTH Machine Design (2005)

15. Karbhari, V.M.: Product and process development methods. JTEC Panel Adv. Manuf. Technol. Polym. Compos. Struct. Jpn. (1994)
16. Lövgren, R.: Product development methods: Leonardo da vinci versus modern engineering education. In: The Renaissance Engineer of Tomorrow: Firenze, 8–11 September 2002: Proceedings of the 30th SEFI Annual Conference. Edizioni Polistampa, Firenze (2002)
17. MEMSIndustryGroup: Technology development process template. http://www.memsindustrygroup.org/i4a/doclibrary/getfile.cfm?doc_id=373 (2012). Accessed 10 March 2013
18. OGC: Managing Successful Projects with PRINCE2, 4th edn. The Stationery Office Ltd, Norwich (2009)
19. Olson, E.L., Bakke, G.: A comparison of new product development practices in North America and Norway. Telektronikk **100**(2), 120–125 (2004)
20. Ortloff, D.: Product engineering for silicon based MEMS IP. PhD Thesis, University of Siegen (2006)
21. Project Management Partners: ISO 10006: risky business. http://www.pmpartners.com/resources/iso10006.html (2005). Accessed 27 May 2013
22. Pugh, S.: Total Design: Integrated Methods for Successful Product Engineering, 1st edn. Addison-Wesley, Reading (1991)
23. Schwalbe, K.: PRINCE2 a Practical Handbook, 4th edn. Kathy Schwalbe LLC, Minneapolis, Minnesota (2013)
24. Siegelaub, Jay: PMBOK and PRINCE2. http://www.slideshare.net/sundong/prince2-vs-pmbok (2005). Accessed 24 January 2013
25. THE WELCH COMPANY: Reviewed ISO 10006 on project management, compared PMBOK. http://www.welchco.com/sd/08/00101/02/95/07/21/134937.HTM (1995). Accessed 24 January 2013
26. Ulrich, K.T.: Product Design and Development, 1st edn. McGraw Hill, Columbus, OH (2003)
27. Verpoorten, J.H.: Projectmanagement – PRINCE2 versus PMBOK. http://www.cs.uu.nl/docs/vakken/pm/sheets/college7_1.pdf (2005). Accessed 22 Aug 2013
28. Wideman, M.: Comparing PRINCE2 with PMBOK. AEW Services, Vancouver (2002)

Chapter 3
Micro- and Nano Systems: A World of Its Own

Abstract This chapter serves as an introduction of the MEMS industry that is the target application area for the methodology presented in this book. The reader will be given a quick overview what MEMS are, how they are made, and who is involved in making them.

3.1 MEMS Products

In the previous chapter some baseline methods for Product Engineering like Stage-Gate® and the newly introduced T^2M have been presented. These methods are applicable to product engineering in almost any field. So why bother focusing on MEMS? What makes MEMS so special that they require an individual approach on product engineering? To find answers to this question we first have to gain an understanding what all this MEMS business is about. In this chapter we will try to sum up what MEMS are, how they are made, and who is involved in making them. First let's start with what MEMS are.

MEMS is an acronym that stands for *M*icro-*E*lectro-*M*echanical-*S*ystem. Basically it can be said that the term describes extremely miniaturized systems with electrical and mechanical components. And in a very strict sense that is exactly what they are. But by looking at their context it becomes clear that they are so much more than just that. Only 20 years ago MEMS were just niche products for very expensive military and automotive applications. Five years ago they were still uncommon. Today there are several of them in every smartphone. So one of the things that makes MEMS so special is their tremendous success that brought them from no-where to almost every-where in just a couple of years. What happened? Looking at probably the most prominent MEMS products, accelerometers and gyroscopes, reveals some interesting facts. In the days before MEMS these things were rather large, expensive, and delicate instruments making their use prohibitive for almost every consumer application—especially for those you are carrying around in your pocket. With the introduction of MEMS this has changed dramatically. Now accelerometers measure

D. Ortloff et al., *MEMS Product Engineering*, DOI 10.1007/978-3-7091-0706-5_3,

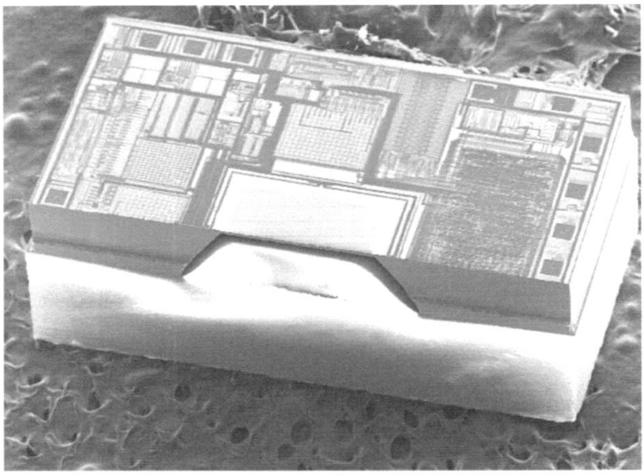

Fig. 3.1 Die photo from an integrated MEMS pressure sensor with permission from Elmos Semiconductor AG

millimetres, weigh milligrams, cost only a few dollars, and don't require regular maintenance.

Figure 3.1 shows a photography of a MEMS pressure sensor made by *Elmos Semiconductor AG*. All mechanical and electrical components of the system have been integrated monolithically on a single piece of silicon. The chip on the photo has been cut in half, so the thin sensor membrane is clearly visible. There have been exactly these three reasons—small size, low cost, and robustness—that made it possible to integrate MEMS in everyday products and as such enabling completely new application scenarios. Especially the areas of mobile applications and location-based services have been turned upside down with the availability of gyroscopes, GPS, cameras, and plenty of computation power in almost every mobile phone.

Despite electro-mechanical devices being the most prominent application area the term MEMS is also commonly applied to micro-devices consisting of components from other engineering domains, e.g., micro-opto-mechnical devices (MOEMS) or micro-fluidic devices. During the last years MEMS sensors have been developed to measure mechanical, optical, thermal, magnetic, and chemical phenomena. Combined with the constantly growing list of MEMS actuators like micro-mirrors to direct light, micro-valves and micro-pumps that enable regulated material transport, and micro-motors and gears that enable positioning on a nano-scale, MEMS have proven to be exceptionally versatile. Books like [3, 4] provide an overview of the huge diversity in MEMS sensor and actor concepts.

Considering that most of the currently available MEMS products integrate only a single sensor or actuator with microelectronic components there is still a lot of innovation potential for more complex systems only by recombining already available components. For instance, MEMS pressure sensors—like the one shown in Fig. 3.1—combined with a MEMS gyroscope for height information and a MEMS compass for absolute direction could serve as an indoor navigation system for large multi-storey buildings where gps-based navigation is not possible.

3.2 MEMS Process Technology

One of the things that distinguishes MEMS products from their classic counterparts and that is to a large degree responsible for their success is the way how they are made. While there are many specific manufacturing techniques like LIGA [1] to cope with the special requirements of applications like bio-implants or micro-gears, the techniques derived from IC-manufacturing have been proven most versatile. Besides the obvious advantage that almost every MEMS needs electronic components these techniques are well established, reliable, and suitable for cost-effective mass-production.

The methodology presented in this book has been tailored to the tasks and processes used to engineer these silicon-based-MEMS products. Therefore this section will give a brief outline on silicon micromachining. Only the most basic process steps and their characteristic parameters are discussed. For a more detailed description of MEMS manufacturing technology, see [5] or [2].

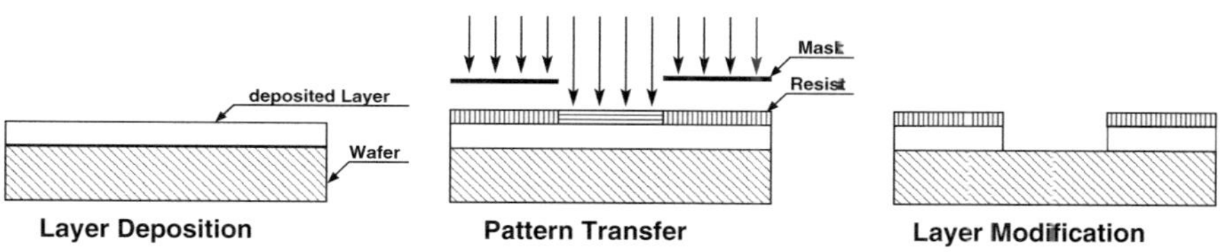

Fig. 3.2 Silicon micromachining base techniques

All silicon-based MEMS have in common that they are built on silicon wafers using photo-lithographic processes. A single wafer can hold hundreds of MEMS. After wafer manufacturing the individual devices are separated, a process called dicing, and put into packages that protect them from environmental influences. Silicon-based MEMS products look just like electronic microchips and can be handled, interfaced, and processed with the same methods and equipment.

The photo-lithographic process used in wafer manufacturing can be boiled down into three base techniques that are applied iteratively to create microstructures on the wafer-surface:

- Layer Deposition
- Pattern Transfer / Lithography
- Layer Modification

Figure 3.2 illustrates the base process. First step is the deposition of a new material layer on the wafer. Next a photo-sensitive material is applied. This so-called photoresist is then exposed through a lithographic mask that holds a two-dimensional geometric pattern. After development of the exposed photoresist the material layer can be selectively removed by layer modification techniques like etching. Applied iteratively with different materials very complex structures can be created on the wafer surface with these steps.

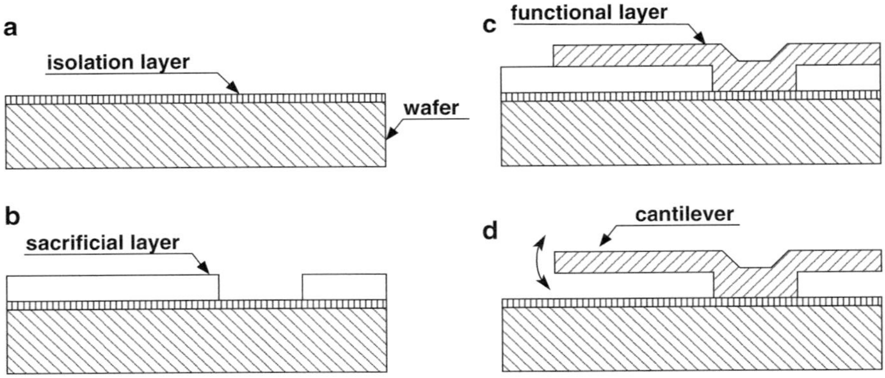

Fig. 3.3 Creating moveable structures using sacrificial etching

One common challenge for MEMS with micro-mechanical components is to create structures that are not rigidly fixed on the wafer surface but that are in some ways moveable like membranes for pressure sensors or swinging masses for acceleration sensors. Figure 3.3 shows how the base process can be modified to create a moveable structure. The trick is to completely remove an underlying material layer, so that the structures above the layer are released from the wafer surface. Because the material layer is sacrificed during the process this technique is also called sacrificial-etching. What looks simple is in fact a very complicated and error-prone process that requires careful selection of materials and application process-steps. For instance, the etching technique has to reliably remove the whole sacrificial layer while not attacking any functional layers. And the moveable structure, once released, must not break, bend, or stick to other structures.

To be able to deposit the many different materials with various thicknesses and properties, to selectively remove material with well-defined speed and direction, and to reliably transfer patterns with high precision onto ragged wafer surfaces with differences in altitude that makes most mountain ranges look flat in comparison, many different variations of the base techniques exist.

3.2.1 Layer Deposition

Layer deposition supplies the MEMS manufacturing process with raw materials. In silicon micromachining material is created either by depositing whole material layers from external sources or by chemically modifying existing material structures on the wafer surface. Typically those techniques are characterized by

- material
- growth rate

- minimum and maximum layer thickness
- process temperature

Layer deposition techniques are usually limited in what kind of *material* they can deposit. And even if two techniques can be used to deposit the same materials, the quality, internal stress, or other properties of the deposited layer may vary. The *growth rate* indicates how fast the layer thickness increases. It is therefore primarily an indicator of process speed but it also influences how long the wafer has to be exposed to the *process temperature* to reach a certain layer thickness. Especially high *process temperature* may also have an impact on other material layers and is therefore itself an important parameter used to characterize layer deposition techniques. Depending on the technique there is also a *minimum and maximum layer thickness* that can be reliably deposited.
The most common deposition techniques are the following ones:

Chemical Vapor Deposition (CVD) is a technique where the wafer surface reacts chemically with the ambient (gaseous) environment. *CVD* is suitable for depositing a large selection of materials. There are several variations with a wide range of process temperatures and deposition rates. *PECVD (Plasma Enhanced CVD)*, for instance, uses plasma to enable lower process temperatures of down to 300 °C while *Low Pressure CVD (LPCVD)* creates layers of better quality and uniform thickness at the cost of a slower growth rate and much higher process temperature starting at about 600 °C.

Physical Vapor Deposition (PVD) uses physical instead of chemical means to deposit material layers. The most prominent *PVD* techniques are high temperature evaporation and plasma-sputtering. During high temperature evaporation the material is evaporated at high temperatures and low pressure. The layer is built up by condensation of evaporated particles on the wafer surface. During a sputter process the material is evaporated by a bombardment of accelerated ions. For sputtering the process temperature is generally lower at the cost of a much slower growth rate. PVD is mostly used for metal deposition.

Electroplating is a galvanic process for depositing metals on a electrically conductive surface by applying an electric current through a solution containing metal ions. The technique is suitable to produce thick layers of conductive materials at high speed and low temperatures.

Oxidation techniques chemically modify the surface instead of depositing materials. For instance, oxidizing a silicon surface creates a layer of silicon-dioxide (SiO_2). This process consumes silicon from the wafer surface, resulting in a reduced silicon height but an increased wafer height because of the additional oxygen atoms. There are two types of oxidation used in silicon micromachining. The first one is *wet oxidation* which is done in an oxidizing atmosphere that contains water vapor. The process temperature is normally between 900 °C and 1000 °C and causes a fast oxidation process. The second type is called *dry oxidation* without the aid of water. Process temperatures are generally higher (ca. 1200 °C). Compared to *wet oxidation* this process produces a superior quality oxide at the cost of a much slower growth rate.

Doping is the process of introducing impurities into a semiconductor material to change its properties (e.g., electrical conductivity or response to a certain etchant). Again, there are two prominent techniques to accomplish this task. During *diffusion* the substrate is put into an environment containing doping atoms in the gas phase. These atoms start to diffuse into the silicon lattice under a high temperature condition (900°C to 1100 °C) because of the concentration gradient between the silicon substrate and the environment. The result is a dopant profile with the highest concentration at the surface of the wafer. *Implantation*, on the other hand, introduces doping material making use of a high energy ion-beam. The accelerated ions are injected into the wafer and lose energy by collisions with the lattice atoms. After a certain distance they have lost most of their kinetic energy and stop traveling. As a result, the highest concentration of doping atoms is not necessarily at the surface of the wafer. *Implantation* requires a much lower process temperature and gives better control of the doping gradient. A drawback is that the crystal lattice of the semiconductor may be damaged by the ion beam.

Spin-on is a technique commonly used to apply polymers on the wafer surface. During this process a liquid material solution is dripped or sprayed on a rotating wafer. The rotation ensures a homogeneous distribution of the material. Layer thickness depends mostly on the material and rotation speed. Deposition is usually performed at room temperature but most materials require a *baking* step at medium to high temperatures to remove the solvent.

3.2.2 *Pattern Transfer/Lithography*

In silicon micromachining the most common technique used to transfer geometric patterns to the wafer surface is photo-lithograpy, which means exposing a layer of photo-sensitive material, called photoresist, through a photo mask that holds a two-dimensional geometric pattern. After exposure the photoresist is placed in a developer solution that will selectively etch away either the exposed or the unexposed region, depending on the photoresist used. Using positive photoresist the exposed regions are removed, using negative photoresist they are resilient to the developer solution. The characteristic parameters of this base technique are

- minimum feature size
- depth of focus

Minimum feature size describes the minimum size of patterns that can be transferred to the wafer surface. In photo-lithography this parameter is limited by the wavelength of the light and by the optical properties of the exposure system used. *Depth of focus* describes the distance range over which the projected patterns are sufficiently sharp. If the photoresist layer is not within this range, the projection is blurred leading to a reduced resolution. This becomes important for exceptionally thick layers of photoresist or if there are large height differences in the topography on the wafer, which may frequently occur in silicon micromachining.

There are three types of exposure systems in classic photo-lithography:

Contact Lithography is the most basic exposure system. In this system the mask is put into direct contact to the wafer. As a result the diffraction losses are minimal and the depth of focus is very high. Despite these ideal parameters there are also some serious drawbacks using contact lithography. The direct contact with the wafer is liable to soil or damage the mask. Another issue is that the mask has to be of the same size and resolution as the wafer. At increasing wafer diameters mask-alignment and the provision of uniform light sources becomes increasingly difficult. For these reasons contact lithography today is mostly limited to research and prototype processing.

Proximity Lithography is similar to contact lithography. The only difference is that a small gap remains between the mask and the wafer. This prevents issues like damaging the mask. Otherwise benefits and shortcomings are almost the same.

Projection Lithography is the system currently in use for high volume production. The wafer is exposed through a projecting optical system. This allows the mask to be manufactured in a much lower resolution than the target system. Therefore dust on the mask is much less an issue as such impurities are also scaled down by the projection system. For MEMS the very narrow depth of focus may become an issue as MEMS can have a very deep surface topography or may use very thick photoresist layers.

Photo-lithographic pattern transfer always requires a photomask and a light source. For very high resolutions or very small volumes the technical and/or financial effort becomes disproportionately large. Therefore alternative techniques like maskless lithography and nano-imprint lithography have been introduced.

Maskless Lithography is applicable in situations in which using a mask is not appropriate (e.g., small volume production). The most common maskless techniques are *electron beam lithography* and *laser direct writing*. Both techniques use a beam to directly write the patterns into the photoresist. Therefore no mask needs to be manufactured and the pattern can be varied for every exposure making those techniques especially interesting for research and prototyping. The gain in flexibility comes at the cost of a much lower throughput, rendering those techniques useless for high volume production.

Nano Imprint Lithography is a technique that transfers patterns by mechanical deformation of the resist. It is suitable for both high volume and high resolution applications. There are many variations, the most common ones being *thermoplastic nanoimprint lithography* and *photo nanoimprint lithography*. Both techniques require a mold for the pattern to be transferred. For *thermoplastic nanoimprint lithography* the mold is pressed onto the resist and heated above the glass temperature of the resist. After cooling down, the mold is removed and the patterned resist remains on the wafer. *Photo nanoimprint lithography* requires a transparent mold that is pressed into a liquid resist. The resist is then cured through the mold using UV-light.

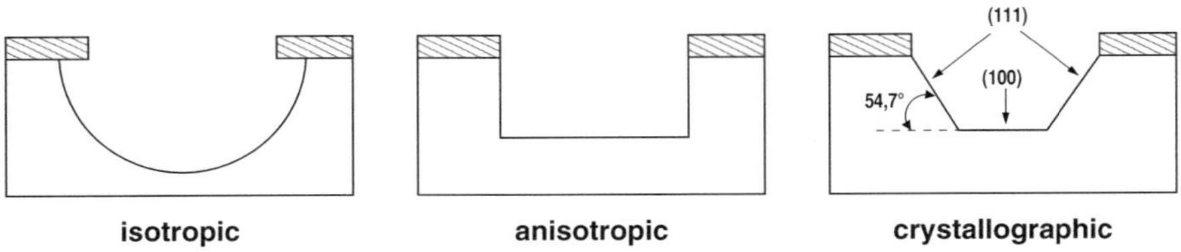

Fig. 3.4 Influence of isotropy on etch profile

3.2.3 Layer Modification

The term *layer modification* is used for all techniques that modify the geometry of a layer by material removal. For silicon micromachining mostly etching techniques are used for this purpose. The relevant parameters are

- etch rate
- isotropy
- selectivity
- maximum aspect-ratio

The etch-rate describes how fast material is removed from the wafer surface. It is therefore a measure for the speed of the modification process. The *isotropy* indicates the direction of the material removal. If a process is isotropic, the etch rate is independent of direction. Most chemical processes belong to this category. For anisotropic processes the etch-rate is significantly higher in a certain direction. For physical etch processes the vertical etch-rate is usually higher than the lateral etch-rate. Figure 3.4 shows how the isotropy influences the profile of the etched shape. *Selectivity* is a parameter that describes the ratio of the etch-rate towards different materials. A high selectivity means that the target material is etched with the maximum etch rate while the other material is mostly unaffected. The aspect-ratio of a layer modification is defined as the ratio between its depth and its width. Therefore deep and narrow trenches require modification techniques with high *maximum aspect-ratio*.

There are many different etch techniques available for silicion micromachining. They are roughly categorized into *wet* and *dry* etching and *chemical* and *physical* etch mechanisms. *Wet etching* uses, as the name implies, liquid etchants. The etch mechanism is almost exclusively chemical. The advantages of chemical wet etching are their high etch rates and their good selectivity. As chemical processes they are in most cases isotropic which usually leads to undercutting and a poor maximum aspect ratio in general. For some chemicals, like *KOH*, silicon is etched anisotropically. Etching in direction of the (111)-plane is considerably slower and leads to pyramid-shaped holes in (100)-wafer surfaces. This kind of etching is therefore also called crystallographic (see Fig. 3.4). It has been exploited in many MEMS designs.

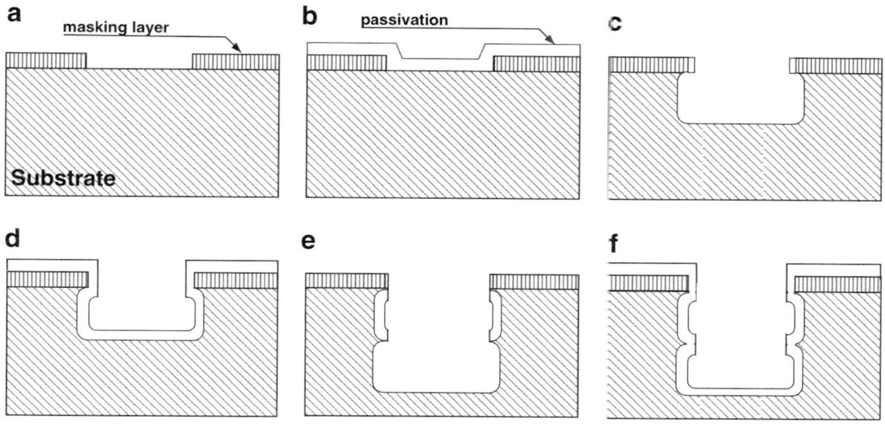

Fig. 3.5 Deep reactive ion etching

Dry etching techniques use gaseous substances and employ either physical or chemical etch mechanisms. Basically there are three categories of dry etching:

Vapor Etching is the most simple dry etching technique. The wafer is exposed to gases that chemically dissolve material at the wafer surface. Like wet-chemical etching this technique shows mostly isotropic behavior and—depending on the materials—good selectivity. Process control is usually better compared to the wet-chemical techniques at similar etch rates.

Sputter Etching is almost the same as the sputtering deposition process. The only difference is that the accelerated ions are used to physically remove material from the wafer surface instead from the sputter target. The same equipment can be used for sputter etching and sputter deposition. Sputter etching is very anisotropic and etches almost vertically. Therefore it can be used for high precision material removal and chemically inert materials. Main drawback is its very low etch rate.

Reactive Ion Etching (RIE) uses a combination of chemical and physical mechanisms. A plasma is added to the reactive gas mixture, creating ions that are accelerated towards the wafer surface. The plasma increases the reactivity of the gas mixture and the accelerated ions add a physical component that is comparable to sputter etching.

Deep Reactive Ion Etching (DRIE), also referred to as *Bosch Process*, is special variant of *RIE* which has become a very important technique for MEMS manufacturing. Using this technique, trenches with very high aspect ratios and almost vertical sidewalls can be achieved. The technique consists of two different process steps that are applied iteratively. The first step creates a polymer on the wafer surface that acts as a passivation layer and the second step is *RIE*. The physical component of the *RIE* removes the passivation at the bottom of the trench, leaving the sidewall-passivation intact. The reactive component now etches isotropically at the bottom of the trench. Before the isotropic undercut becomes too large the sidewalls are passivated again and so on. Figure 3.5 illustrates the process.

3.3 Business Models in the MEMS Industry

In the first part of this chapter it became clear that MEMS Product Engineering is a highly interdisciplinary effort requiring a mixture of specialist knowledge and technological and logistical capabilities that depend to a large degree on the envisioned product. Considering the variety of application areas, technologies, and manufacturing capabilities, it is very unlikely that a single company can cover all aspects of MEMS product engineering. A cooperative approach is much more likely. And indeed a recent analysis of the MEMS market [6] shows several emerging business models that cover only parts of the value chain:

MEMS Device Development is the first step and deals with MEMS development on the system level. Here the overall functionality of the system and its components are defined.

MEMS Design is the step where the physical implementation of the components is designed, e.g., the micro-mechanical components and the microelectronic circuits.

Technology Development is the task where the manifold manufacturing techniques described in the last section have to be combined into a process that is able to manufacture the MEMS device. This is arguably the most time-consuming part of MEMS product engineering.

MEMS Wafer Manufacturing describes the process of manufacturing the devices on silicon wafers. In some cases the micro-mechanical and microelectronic components are manufactured separately on different wafers.

MEMS Assembly and Test is the part of the manufacturing process where the wafers are diced into individual chips, assembled into packages and tested. For several reasons these steps are specific to the MEMS and IC-industry this is a separate step that is often performed at other places and by other companies than the wafer manufacturing.

Marketing and Sales finally is the step where the manufactured product is brought to the customer. This is usually the domain of local distributors.

Figure 3.6 contains a selection of established and emerging business models and their position within the MEMS value chain. The classic business model that covers the complete value chain is that of the integrated device manufacturer (IDM). Besides the aforementioned disadvantages that led to the more specialized business models this model still benefits from having all competencies and responsibilities bundled in a single company. In the group of emerging business models we see two complementary models that have proven very successful in the microelectronics market: the *fabless design house* and the *pure play foundry*.

The *fabless design house* business model concentrates on system design challenges and relies on external foundries for manufacturing. Main benefit of this business model is, on the one hand, that it avoids the huge investments necessary to build up fabrication capacities, thus allowing also small companies to enter the market. On the other hand the lack of own fabrication sites gives the *fabless design*

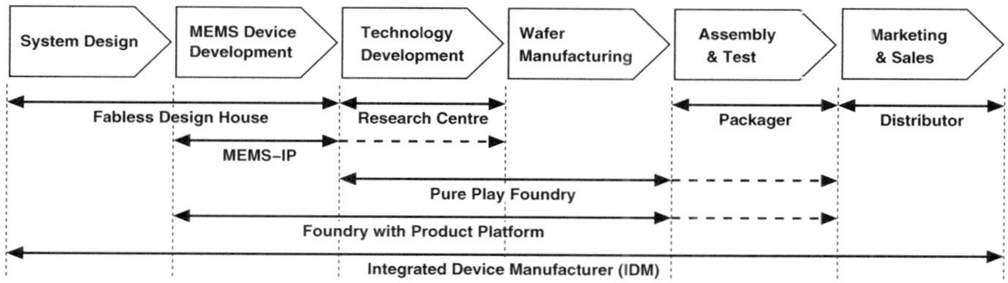

Fig. 3.6 Business models along the MEMS value chain

house the freedom to select a manufacturing partner based on the individual design project.

Companies maintaining manufacturing sites have the problem that the huge cost to operate such a manufacturing site requires a very high utilization to keep it profitable. One possible way to increase utilization is to offer surplus manufacturing capacities to other companies. The *pure play foundry* business model is to consequently offer all manufacturing capacities to customers. This allows them to concentrate fully on the development and optimization of their manufacturing technologies. By deliberately not developing MEMS products, they are not seen as competitors by the design houses making it easier to acquire contracts.

As most MEMS companies are or have been active in the microelectronics markets, the emergence of these two business models came almost naturally. However the clear interfaces between design and manufacturing that made them successful in the IC-industry are not directly applicable to MEMS. There are strong interdependencies between device design and manufacturing technology that cannot be broken up easily. Offering pure MEMS manufacturing processes requires a very intensive cooperation with the design house. Two emerging business models that address this problem are the *MEMS-IP design house* and the *foundry with product platform*.

The *MEMS-IP design house* combines MEMS design and technology development to offer MEMS-IP (intellectual property). While still as fabless as the *fabless design house* the *MEMS-IP design house* needs a laboratory to develop their processes and close cooperation with manufacturing partners to be able to deploy the processes. For the last reason the *foundry with product platform* emerged. In contrast to the *pure play foundry* this business model offers additional MEMS design services on the MEMS-IP level. In such a design scenario the *fabless design house* still designs the overall product and its microelectronic components but leaves the design of MEMS components to the *foundry with product platform*. The critical interface between MEMS-design and manufacturing is kept within one company. To be able to offer such services the company has to invest in developing MEMS-IP—the so-called product platform—for their processes.

Other typical business models in the value chain are *research centers* that offer advance development of new fabrication techniques and *packagers* and *distributors* that extended their services from the IC-industry to MEMS. In summary we see two main parties involved in the MEMS Product Engineering business:

- the developers that develop the device components and the manufacturing process,
- the manufacturers that manufacture, assemble, and test the devices.

What is still missing is the driving force behind the product engineering process that has the product idea and an interest in actually making a real product out of it. For standard products a *design house* or a *distributor* might take over this role but as we learned in Sect. 3.1 MEMS products are often highly application-specific and are not marketed as individual products but as parts of other products like mobile-phones. In this case the manufacturer of the final product—the customer of the MEMS value chain—would be the driving force behind the MEMS product engineering. Besides the financial interest, the customer is the only party that has intimate knowledge of the target market and the constraints of the application area. These are the reasons why the methodology introduced in this book puts the customer in charge of the product engineering process.

References

1. Becker, E.W., Ehrfeld, W., Mnchmeyer, D., Betz, H., Heuberger, A., Pongratz, S., Glashauser, W., Michel, H.J., Siemens, R.: Production of separation-nozzle systems for uranium enrichment by a combination of x-ray lithography and galvanoplastics. Naturwissenschaften **69**, 520–523 (1982)
2. Franssila, S.: Introduction to Microfabrication. Wiley, Hoboken (2010)
3. Gad-El-Hak, M.: MEMS: Applications. CRC Press/Taylor & Francis, Boca Raton (2006)
4. Korvink, J., Paul, O.: MEMS: A Practical Guide to Design, Analysis and Applications. William Andrew, Inc., Norwich (2005)
5. Madou, M.: Fundamentals of Microfabrication, 2nd edn. CRC, Boca Raton (2002)
6. Yole Développement: Status of the MEMS Industry. Yole Développement, Lyon-Villeurbanne (2010)

Part II
MEMS Product Engineering Aspects and Methods

This part forms the core of this book. Chapter 4 explains the setting addressed by the methodology. Here the authors' views on MEMS Product Engineering are discussed in detail. The reader is introduced to the various tasks that have to be performed during Product Engineering, including device design, technology design, and quality management. Here, also the general approach of putting the customer in charge of the Product Engineering process is introduced.

Chapter 5 is the heart of this book and describes the methodology designed to support distributed MNT Product Engineering. The chapter gives an overview of the method composition and then dives deep down into the details of the individual methodology blocks. It follows through the methodology in chronological form starting from the first product ideas to the product entering the market giving the details of the required business processes.

The last chapter of Part II of this book bridges the abstract methodology to a more practical application of it. It introduces the most critical infrastructure components because the methodology requires significant tool support for its processes to be efficient. Additionally these tools need to provide certain features that are required for a sufficient MEMS Product Engineering Infrastructure. Chapter 6 describes the most essential features extracted during studying such a business environment.

Chapter 4
MEMS Product Engineering

Abstract This chapter points out the individual challenges of micro electro mechanical system (MEMS) design issues. Starting from device design according to the idea of the product a concurrent technology design (determining the manufacturing flow) is necessary because of the strong dependencies between structural and technological properties. Additionally the challenges of packaging and assembling have to be considered in early design stages. Last but not least the aspect of quality assurance becomes more and more important. Modern strategies like Design for Six Sigma (DfSS) adapted to the MEMS design area can substantially improve the quality and the yield in later production processes.

4.1 Getting Started

Product Engineering (PE) has already been well defined in earlier sections of this book. In this chapter the challenges of PE in the area of MEMS are within the focus. It differs substantially from other PE approaches. Even very close technology areas like IC design and fabrication use completely different procedures. For instance, Product Engineering in the field of (VLSI) microelectronics mostly deals with the aspect of *managing complexity* since a set of only few basic components (transistors, diodes, etc.) multiplied in million instances can realize a huge variety of electronic systems [20]. For the field of micro- and nanotechnology (MNT) the main theme could be summed up as *managing diversity*. First there is diversity regarding the application areas that cover the automotive, medical, and consumer electronics sector among many others. Each area has very different technical and market constraints. Second there is a diversity in engineering domains. MEMS products often combine microelectronic, micromechanic, microoptic, and microfluidic components. Each has different approaches to design and test. Third there is a diversity concerning the available manufacturing technologies. This is best expressed by the MEMS law that states that every device needs an individually

D. Ortloff et al., *MEMS Product Engineering*, DOI 10.1007/978-3-7091-0706-5_4,

tailored manufacturing process. Finally, there is diversity regarding the business models of companies involved in MEMS Product Engineering covering the complete scope from *Integrated Device Manufacturers* (IDM) to *Small and Medium Sized Enterprises* (SME) for only limited tasks.

Recent market surveys like [24, 25] indicate that, besides some large IDMs, there are many SMEs that cover only small parts of the value chain (see Fig. 1.1). In such a scenario Product Engineering is a collaborative effort, comprising several specialized companies that carry out different development tasks at different locations. Most prominent example is the distinction between fabless design houses and pure-play foundries which is already common in the IC industry and is now becoming relevant for MEMS as well [24]. Developing a PE methodology that supports such a diverse and distributed scenario depends on a detailed analysis of the mechanisms that currently drive Product Engineering in this area. One aspect is that typical MNT products like acceleration sensors, gyroscopes, ink-jet heads, or MEMS microphones are not stand-alone products. They are components in more complex systems (e.g., smart phones, cars, and medical appliances) with the individual requirements of each MNT product dictated by the application area of the complex system. Therefore the system design approach and technology selection need to be adapted to those system requirements. On the other hand there is a strong influence of the manufacturing technology on the Product Engineering process. MEMS PE is more often than not driven by technological capabilities instead of market needs. The result in this case is more a technology demonstrator than a marketable product. If the goal is a commercially successful product, the engineering process needs to be market driven instead of technology driven.

4.2 Device Design

Device design from system level to structural level for MEMS generally takes important decisions since this stage not only influences the structure of the product but (unlike in microelectronics) determines strongly also the technological requirements for the manufacturing process. For silicon-based VLSI, in contrast, the chip design benefits from a clear interface between manufacturing and design. This is possible due to some characteristics of integrated circuits [1].

Rectangular polygon-based design, fixed process flows, and shape-independent behavior of the (electrical) components allow a reduction of the structural design to 2.5 dimensions (planar layout plus technology constraints delivered as geometric design rules). In combination with the Y-chart model of Gajski et al. [4] this so-called Mead–Conway abstraction forms the basis of nearly all design concepts and design automation in the field of microelectronics [14].

With the challenges offered by deep submicron, mixed-signal and also mixed-domain technologies that reduction is already problematic and for technologies like MEMS or NEMS (nano-electro mechanical systems) this approach is not

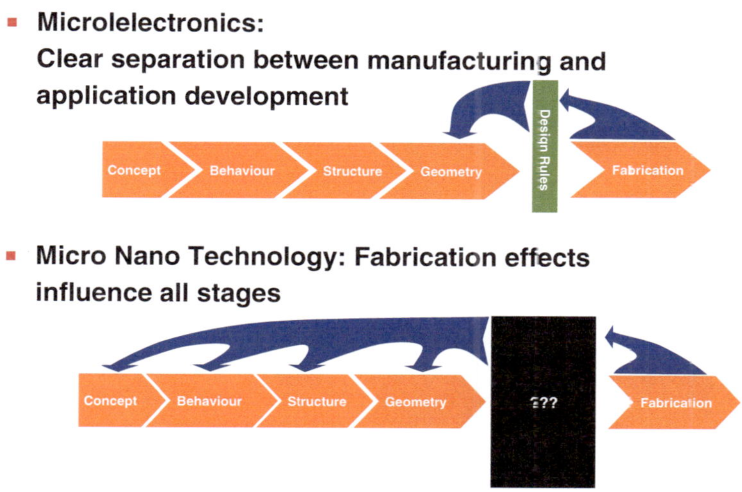

Fig. 4.1 Manufacturing influences in electronics and MEMS design

directly applicable. The Mead–Conway abstraction can hardly be adapted to those emerging technologies because it cannot take into account technological constraints individually tailored towards the requirements of a single device.

In the MEMS area the overall design flow is clearly dominated by manufacturing aspects. Changes at any level of design have an impact on fabrication and vice versa. The three-dimensional structure cannot be reduced to a two-dimensional representation since the third dimension is subject to the design space. Additionally fabrication processes are not fixed but vary from design to design. At the physical design level the device geometry meets the manufacturing process and the material properties. The three-dimensional geometry and the material properties that are mainly responsible for the functional behavior of the device depend mostly on the capabilities and constraints of the manufacturing process. That often leads to the development of application-specific processes (MEMS law) which also means new design rules and technology files for virtually every design. These impacts and the difference to the pure electronics design process is depicted in Fig. 4.1. Therefore the engineering of new MEMS products demands direct access to comprehensive technology knowledge.

A short-term solution to handle the dependencies between design and manufacturing aspects could be a structured physical design flow as proposed by Brück and Hahn [6, 21] in a circle model as shown in Fig. 4.2.

The circle model describes an iterative design flow implementing design, validation, and redesign. Such a physical design flow is currently very common in the MEMS community. The disadvantage of this model is that it concentrates on the physical design without including behavioral design aspects. Newer methodologies combine behavioral and physical approaches to a new concurrent design flow.

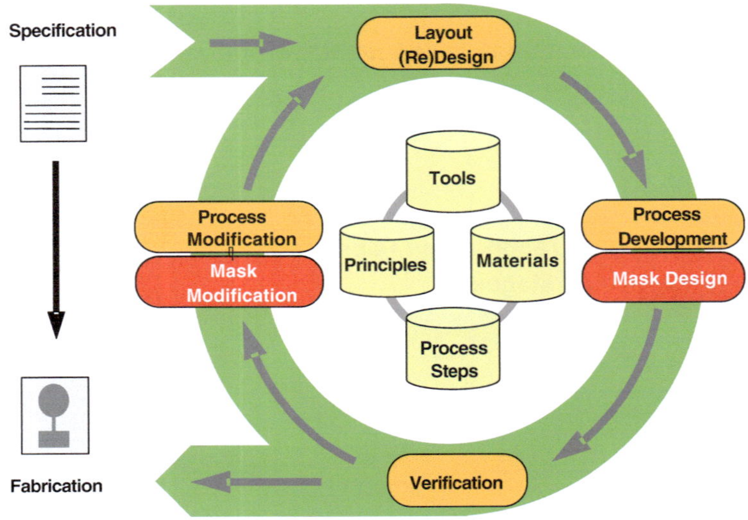

Fig. 4.2 Circle model for MEMS design flow [6]

The actual design flow (e.g., for new mixed-domain devices) starts with the physical aspects, proceeds to field modeling and subsequently to macro-model generation and to behavioral modeling. The data needed for modeling has to be collected and prepared manually. The available design tools cover at least three different Product Engineering domains:

- specialized design tools for the behavioral approach with support of, e.g., macro modeling, mask layout, and FEM simulation.
- tools related to technology aspects (TCAD tools) like process assembly and simulation
- technology design tools

To support the implemented design flow, there is a strong need for technology support with interfaces to behavioral modeling. Such an interface has to provide means to hand over and prepare data from technology design to system and field modeling. A possible design flow for a complete new device with unknown process flow could be as laid out in Fig. 4.3.

This so-called "pretzel" model illustrates the possible flow: The designer starts with creating a structural description based on the requirements (left hand side). A corresponding 3D-model of the device can be derived from the structural description. Now a process sequence for manufacturing the device can be designed (right hand side). Additionally important material and process step data is collected and passed to the knowledge base. The process sequence can now be verified by

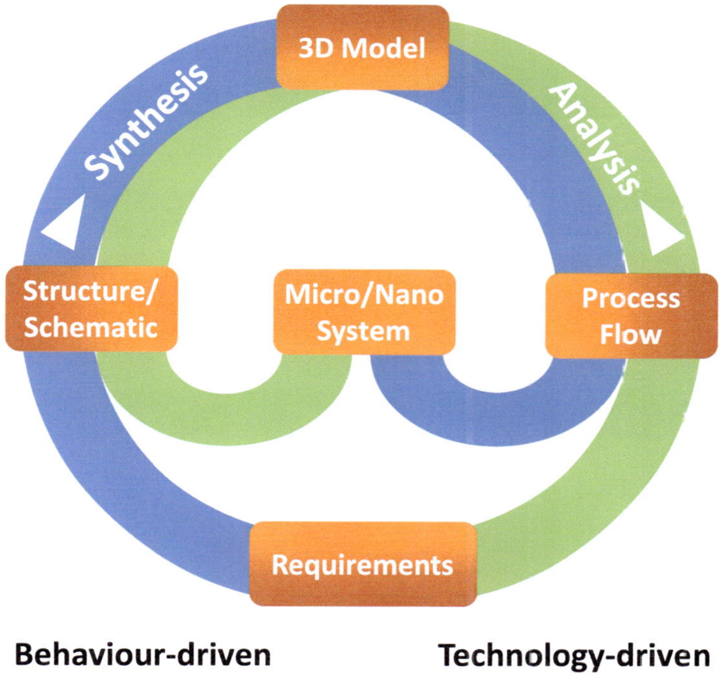

Fig. 4.3 Pretzel model for MEMS design [10]

simulation based on the data in the knowledge base. The steps have to be repeated until the process sequence matches the 3D model. With a growing knowledge base the concurrent design becomes more and more straightforward since less design cycles are necessary. To realize this design flow in terms of design tools means to manage the manufacturing relevant data and to extract the inherent knowledge for modeling. In fact there is a need for a comprehensive and sophisticated process management environment, which provides evaluated data for behavioral modeling. The process design execution described in the next section comprises the tasks necessary to conduct the bottom-up approach as shown in right-hand side of the pretzel model. Within the Product Engineering scenario it supports the manufacturing part of the complete product engineering cycle. The final step of the analysis flow is the comparison of the structural model with the device prototype (e.g., by test patterns). As a consequence the left part of the pretzel covers behavioral aspects whereas the right part is dominated by fabrication and technology aspects. The model does not imply when or in what order the design steps have to be performed. Similar to the Y-model many different design flows can be devised.

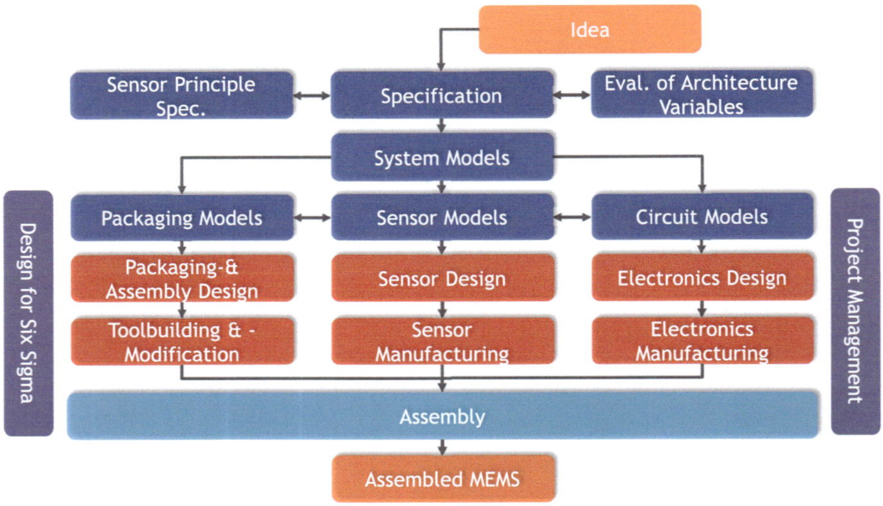

Fig. 4.4 Concurrent MEMS design

4.2.1 Systems Design

For MEMS the entire Product Engineering procedure can be divided into three sectors comprising the development of:

- microelectronic components
- non-electrical components (not only mostly sensors but also actuators, etc.)
- an appropriate mounting and packaging environment

All development activities in the three sectors may happen concurrently but must be controlled by a supervising project management and a common quality assurance. For the latter task the "Design for Six Sigma" (DfSS) methodology [23] has turned out to be most promising in the recent years. An introduction is given in Sect. 4.4.1.

Figure 4.4 shows the complete design process starting from the specification and system modeling to the three concurrent strains of package design, sensor design, electronics design to assembly and packaging. It is of importance that the basic considerations for package design, sensor design, and electronics design are taken into account on high levels to avoid loopbacks in the design flow caused by incompatibilities or the disregard of restrictions from individual design domains in the early stages.

In either way on system level the MEMS design task starts in general with the idea of the customer who often wants to integrate the MEMS device in an end user system. Normally that customer is not the end user but a "tier one" or "tier two" supplier in the entire value chain. Tier one reflects in this context the Original Equipment Manufacturer (OEM) whereas tier two relates to the component manufacturers. The key functionality and the main restrictions with regard to materials, sizes, and cost are direct outcomes of customer's guidelines. In the preparation phase the requirements will be analyzed and the feasibility will be checked. The fulfillment of customer as well as market requirements is in the focus of this stage. The successful engineering of MEMS products requires the ability to understand the customer's needs. The implementation of this kind of customer orientation in the Product Engineering process can be realized by using the "voice of the customer" methodology. This approach tries to capture and structure all customer requirements in order to have the highest degree of clearness of the priority of customer's needs. DfSS provides the guidelines and methods for this kind of procedure. The methodology is sketched in [23].

The initial MEMS design stages can be divided into five sections separated by four gates as shown in Fig. 4.5. The result of the first gate is a formal requirement specification reflecting the needs of the customer. The main product characteristics are clearly highlighted and the feasibility is checked. Additionally qualification criteria and test possibilities are identified at this stage.

If all prerequisites are met, the project partners can launch the next stage resulting in a formal requirements specification (gate 2). Again DfSS methods can define the procedure. This results in a more structured requirements specification containing and highlighting the critical design parameters.

These initial stages of the Product Engineering process are well defined and can be clearly described in engineering guidelines. The next stages defining the system design are much more application-specific and depend on product type, reusability, and previous knowledge of the designer.

In this stage the system will be described and simulated by behavioral models. Certain simulation programs as well as multi-domain simulators and hardware description languages like Verilog-AMS or VHDL-AMS will be used. The result is an interim system specification and a partitioning to the individual domains.

After sign-off the work within the domains (electronics, sensor engineering, assembly design) can start. Time and resource management can be adjusted and the project plan can be adapted.

The following Fig. 4.5 summarizes the initial design stages

4.2.2 Electronics Design

Now the initial stage of the electronics development starts resulting in a floorplan. This part is illustrated in Fig. 4.5. Electronics design is fairly well described in many publications. A good reference for the physical IC design is given, e.g., in [13]. Therefore in this chapter only the major points are highlighted. Circuit ideas are described by models (ideal, behavior, macro, fault, components) or by numerical, analytical or analog / digital simulation. The suitability is checked and optimized. The goal is a robust design that allows the reliable production even in case of disturbances such as production fluctuations and temperature changes.

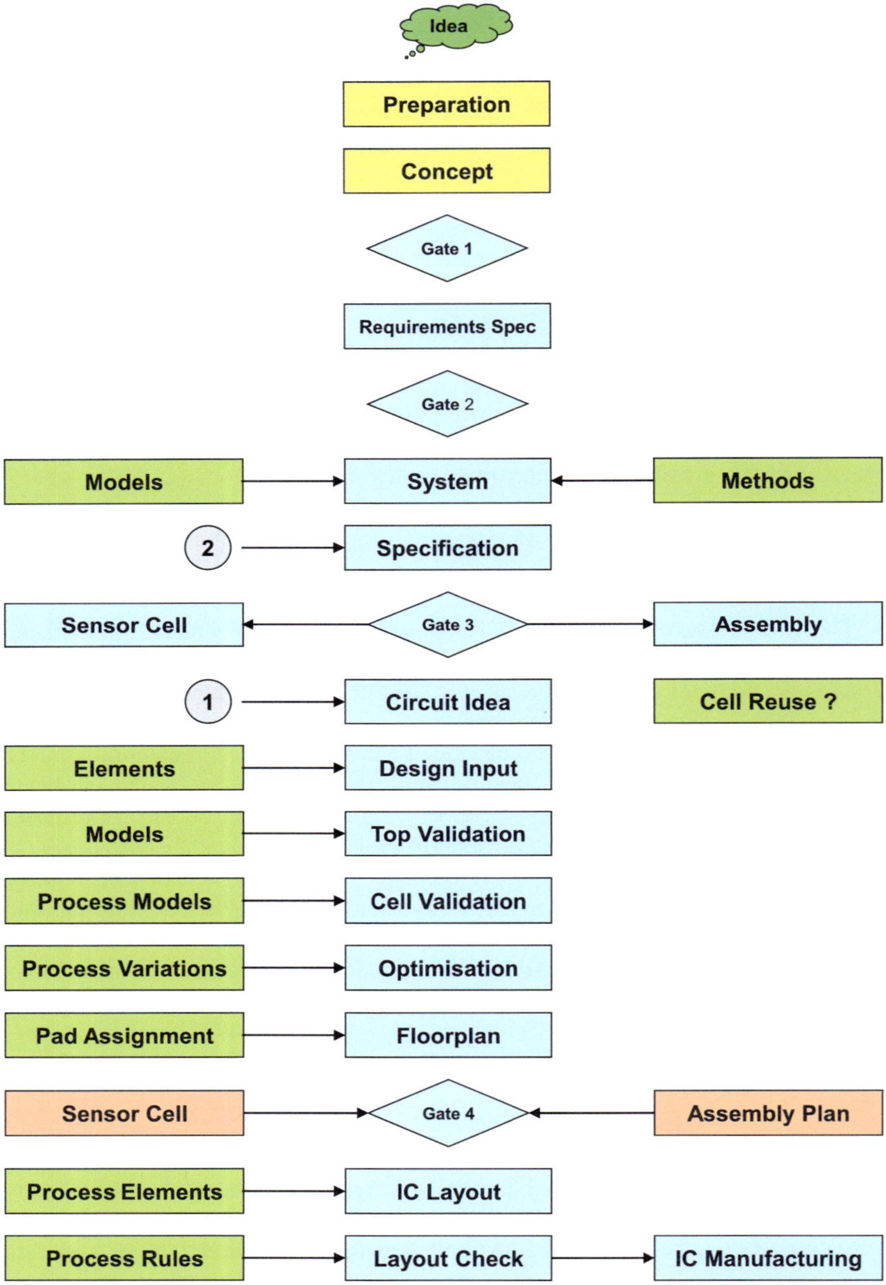

Fig. 4.5 Initial design stages for MEMS design [7]

For quality assurance in an electronic device development many parameters have to be controlled and optimized. These processes are very time consuming and must be automated as far as possible. For electronic design in conjunction with MEMS, the focus is clearly on the analog electronics. The digital portion in many cases is covered by a processor. The simulation is often carried out by tools with simulation engines using equations or hardware description languages like VHDL / Verilog. On completion of the development is Gate 4 (Design Review). Here the separate developments started at Gate 3 will be merged. The interaction of electronics with the sensor and mounting / packaging development is reviewed and compliance with the system specifications is monitored.

The further development stage begins with the IC layout. The layout of analog parts is still much manual work. Due to many constraints, automation is very difficult. To some extent place and route or auto routing are used. The layout checks (DRC: Design Rule Check, LVS: Layout Versus Schematic, ERC: Electrical Rule Check) can be automated by batch processes. The detection of parasitics within the layout is becoming increasingly important to meet, e.g., timing constraints. With the completion of the inspection activities production documents are generated. As already mentioned, also an at least partial automation is very important. At Gate 5 (see Fig. 4.6), the system check is finished and the production of each checked part can be released [7].

4.2.3 Sensor Design

Sensors (or less often actuators) are developed concurrently with the electronic parts. Unlike in microelectronics the design procedure has to reflect the strong interdependency between structure and manufacturing process steps. The circle model as well as the pretzel model takes these design constraints into account. Additionally the mounting and assembly stages have to be developed in order to ensure the packaging of the entire system at the end of the PE process.

As depicted in Fig. 4.7 the details of the sensor design start with the sensor idea defining the functional principle. In many cases more than one sensing principle is available for use. The selection of the most appropriate sensing principle is a crucial part in the sensor design as it determines all subsequent design steps and influences quality and cost. In the case of, e.g., a pressure sensor for vacuum pressures could be realized by applying at least two different sensor principles:

- a thermal sensor principle, in which the gas from the pressure-dependent cooling of a thin film membrane is detected by thermal sensors (Pirani sensor)
- a piezo-resistive sensor principle, where the pressure-dependent deflection of a diaphragm is detected by piezo-resistive layers.

Starting from the sensor principle the design specifications can be developed. In our example the two specifications of a Pirani sensor, on the one hand, and a piezo-resistive sensor, on the other hand, have emerged:

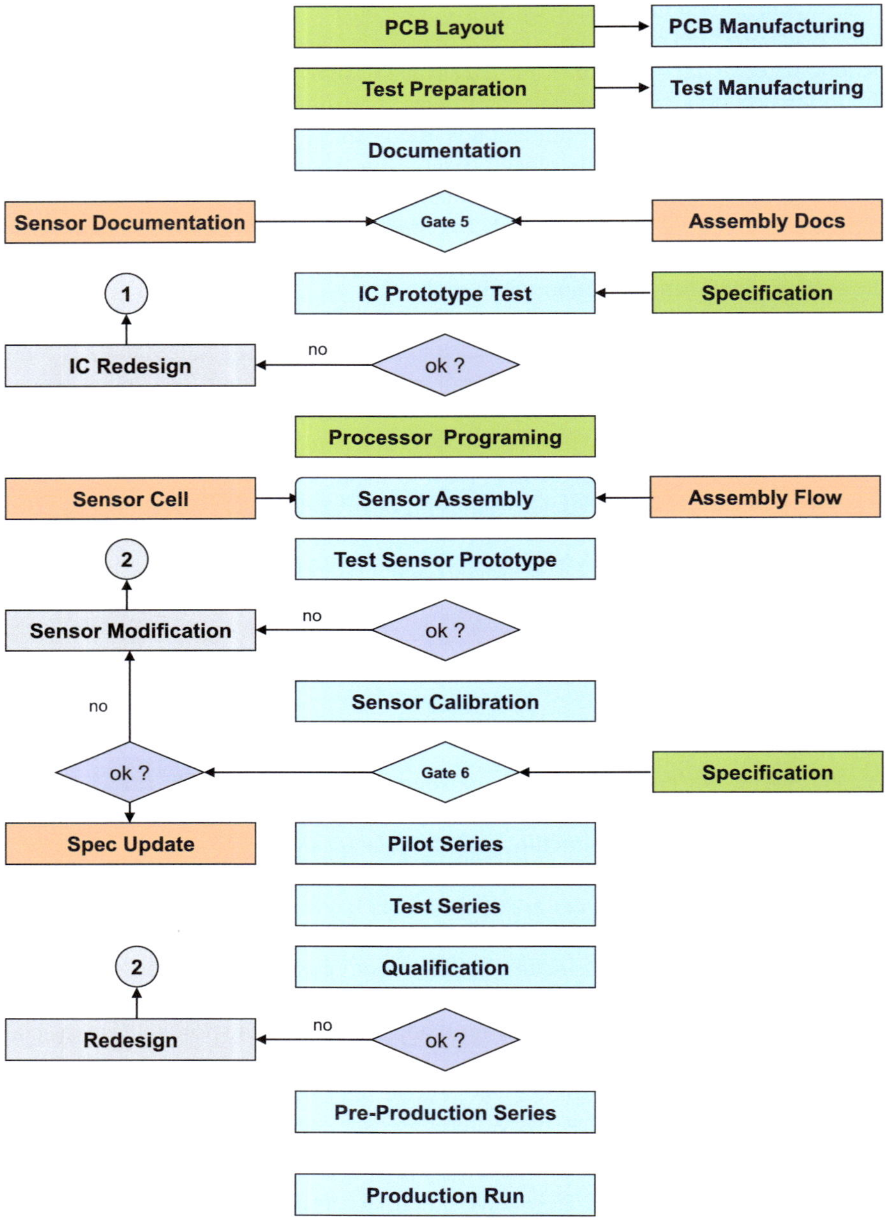

Fig. 4.6 Completing design stages for MEMS design [7]

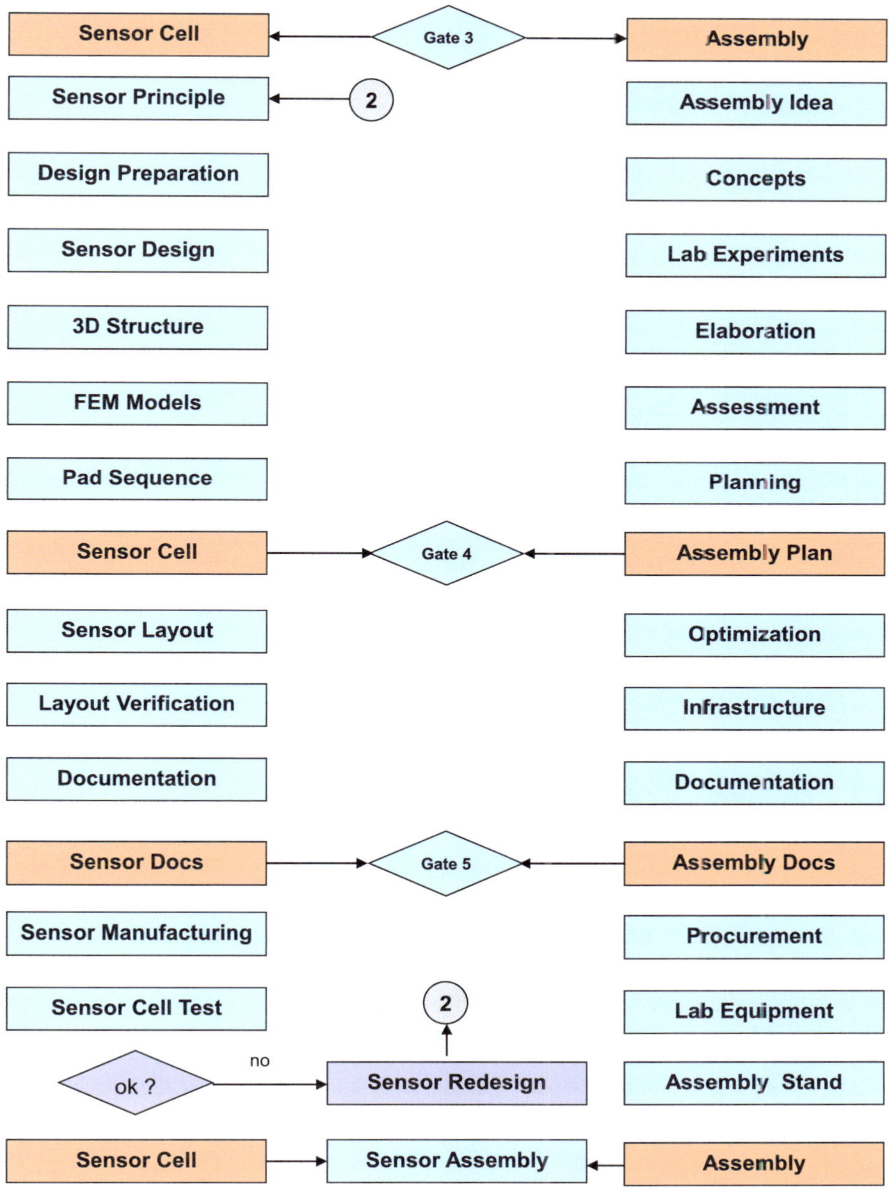

Fig. 4.7 Design steps for sensor elements[7]

4.2.3.1 3D Structure

In most cases the specification and additional calculations already determine the structural composition of a sensor. In our example the calculations in the thermal conductivity model may show that the desired sensitivity at very low pressures below 10^{-4} mbar can only be achieved with self-supporting membranes ("suspended membranes") as a carrier for the thermally sensitive resistor layers.

4.2.3.2 FEM Simulations

The 3D structure will often be simulated before manufacturing steps take place since structural simulation results can validate design assumptions. Simulations are also used to optimize design parameters. Finite elements method (FEM) is a numerical method for solving partial differential equations. It is a widespread calculation method in engineering and is the standard tool for the solid simulation. Simulations based on FEM are used to calculate various properties in solids like temperature, pressure, or stiffness. For the vacuum sensor example the result of FEM simulation, e.g., shows the temperature distribution in the sensor chip in dependence on the respective pressure. This result is valid for the chosen design with the actual specifications defined. Thus it is possible to describe the expected sensor behavior as a function of pressure and to carry out an evaluation of the specified sensitivity [7]. A comprehensive view on the design steps for sensor elements is given in Fig. 4.7.

4.3 Technology Design

Technology design in the MEMS area refers to the task of configuring an appropriate product-specific manufacturing environment. Since the MEMS law (*one product—one process*) applies, the (manufacturing) process development turns out to be a central part of the technology design. Development projects for new manufacturing recipes are challenged by a variety of different external and internal requirements and constraints. The diversity of technology options and their constraints as well as the shrinking geometries and other external and internal forces put the engineers and technologies to their limits. To cope with this growing complexity in fabrication process development new approaches for adequate process design automation cover the design of new process sequences from the very first ideas to the final handover to mass production.

Every new product or product enhancement starts with a new idea. Experience gained by previous developments, scientific papers, and old lab-books provide the major contribution to the realization of these new ideas. Some problems are already encountered at these first stages of the product engineering:

- Colleagues are not always available and lab-books are all too often of value only to the persons who wrote them.

- Even if the information is available in computer files, they are in most cases distributed on several file servers or hidden in places where nobody will probably find them.
- Research papers can provide ideas, but the essential part of the data is missing in most cases.

So already in these first phases of process developments good ideas are scrapped simply because there is no time to perform the necessary research within the information sources and knowledge bases.

However, this is only the beginning of the engineering endeavor. Adding external obstacles like market forces, insufficient internal information, knowledge management, limited development transparency, restricted virtual prototyping possibilities, and missing electronic exchange make many development activities daunting and expensive tasks. This leads to a repetition of experiments, which is costly in terms of time, resources, and money. Experts in semiconductor process development estimate that 10–15 % of failed experiments could be prevented, if previous results would be accessible in an easier way. Furthermore in [2] the analyst company IDC comes to the conclusion that a staggering number of 40 % of R&D experiments are repeated at least once.

4.3.1 *Process Development Methodology*

In the targeted environment of MEMS product development the methodology to develop the process technology is of major importance. To support the process development efforts one structured approach will be documented in the current section. This process development method is integrated as a subprocess into the overall development method, described in Chap. 5. The method requires software support to provide for efficient and reproducible development. Such a software environment and its steps and support is briefly summarized in Sect. 4.3.2. In summary this solution provides support for the tasks of experimental verification integrated into an environment supporting *Design For Manufacturability* (DFM) and process simulation. This environment is described in more detail in the publications [19, 21, 22]. It is suited to support the proposed process development methodology described in following sections.

4.3.1.1 Demonstrator Approach

Analyzing different fabrication lines it can be concluded that each line uses different equipment or slightly different calibration of the same equipment. For microelectronic fabrication these deviations are less of an issue but for the manufacturing of silicon-based MEMS these differences have a high impact. This is due to the additional mechanical requirements. MNT product development becomes even more complex due to the potentially large property changes of the "same" material after porting and integrating a process module at a customer site. In process development for a distributed supply chain it is more realistic to use a *"product demonstrator"* approach, because the process (sub-)flow integration and implementation (materials, machines, etc.) is unknown. This approach is described in more detail below.

The *product demonstrator* approach is based on the following ideas:

- Develop a *product demonstrator* showing that it is feasable on silicon in an industry line or an industry-based institute and generate the knowledge how to use this demonstrator.
- Evaluate results based on the *product demonstrator* for the areas of device and chip characterization, determination of process window, device and chip yield and device and chip reliability data.
- Build up knowledge around the *product demonstrator* (cause–effect chains, impacts of parameters, what-if, etc.) which should help to speed up the re-development and knowledge transfer during the conveying of the product to a customer line.

Within the demonstrator approach the achieved solution can be considered as a single point or line solution where only exactly this recipe in this environment works. This solution can be presented as a proof of concept to the customer. Variations of the process recipe are not yet tested and assessed or only to a very limited degree. To make the solution more robust, the process window for each process step needs to be widened to allow for variations at a location the process is ported to. The characterization for yield, reliability, lifetime, endurance, etc. of devices manufactured with the single line process can be performed in parallel with the widening efforts. Within the product engineering method described in Chap. 5, the demonstrator prototype (single line solution) is developed within the Prototype Development Stage (Sect. 5.5.1) and the characterization and further developments are performed in the Product Development Stage described in Sect. 5.5.3.

4.3.1.2 Subsequent Development Phases

The overall process development is performed in two phases. The first phase develops the above described single line solution exploring only a relatively small set of possible combinations of degrees of freedom. Figure 4.8 presents the degrees of freedom or dimensions as colored boxes. For each of these dimension a small set of discrete points and their combinations are built and tested during the first phase of the development. These discrete points are shown as the fan-outs in the colored boxes in the figure. This first phase develops in the depth, finding a single point or very narrow set of solution points for every dimension of the solution space, i.e., geometrical (width, length, thicknesses), or process driven (flow rates, materials, etc.). The approach implies that in this first phase only a small set of paths will be developed until a working *product demonstrator* is achieved.

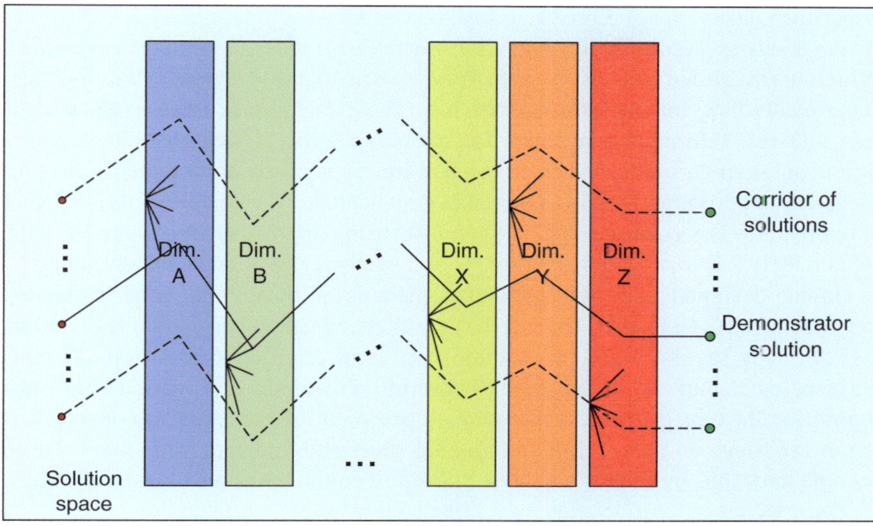

Fig. 4.8 Process development methodology

The knowledge around the single demonstrator solution or small set is built during an extension phase. This second phase widens the multiple paths of the initial solution into a corridor by exploring interesting dimensions further and comparing the results against the already achieved working demonstrator. For that, each dimension of the solution space (degree of freedom), e.g., single material, is explored into the width to build up a bandwidth or process window for each of the possible and sense full dimensions. This adds to the knowledge base captured by a software system as described below in Sect. 4.3.2. Wider process windows enable better portability and increase the number of potential customers. The achieved discrete points of each dimension of the solution space are considered as a bandwidth (process window) with minimum and maximum boundaries. Potentially there might be some bandgaps implying that simple interpolation for the solution points might not be feasible in all cases.

4.3.1.3 Process Window

To perform the width exploration of each feasible dimension in the second phase of the development, it is necessary to know the dependencies between the different possible settings for a single step. This requires to have inspected the cause-and-effect dependencies between the different degrees of freedom. To evaluate the cause-and-effect dependencies, a Design of Experiment (DoE) approach can be used. DoEs, summarized in, e.g., [12] and described in more detail in [18], extract the major dominating parameters concerning a set of effects. Design of

Experiment takes care of reproducibility and unnecessary repetition issues as well as coverage issues. Additionally some DoE methods provide for strategic experiment reduction (fractional factorial design) to extract the required dependencies with less experiments than full factorial approaches. These techniques trade in knowledge gain and full information retrieval for efficiency gains. Using these techniques, significant savings in time and effort can be gained because of the reduced number of necessary experiments to extract the main dependencies. To be able to use this kind of approaches it is essential to plan the efforts using appropriate tools like MiniTab [15] or JMP [11] to be able to cautiously and strategically plan the experiments.

Having designed a set of experiments, the amount of physical experimentation can be reduced even further. To limit the overall amount of experimental verification, the following section highlights a three-step verification approach, first checking the principle manufacturability, then performing process simulations and afterwards supporting the experimental verification. A precondition for using this approach is to put the focus on conserving and making accessible the acquired knowledge to become traceable and reproducible and have full control over the knowledge base.

The extracted dependencies and gained knowledge can be used to adapt the process parameters and to widen the process window getting a more robust process step or entire dimension. Combining all successful experiments a corridor of knowledge can be built. This scenario is sketched in Fig. 4.8 where a corridor of possible solutions is built around the single line prototype solution.

4.3.2 Technology Design Methods and Software Support

Trying to configure a feasible manufacturing solution, it becomes clear that most problems are caused by the large number of parameters. When the semiconductor industry faced the problem of increasing layout complexity, the usage of Electronic Design Automation (*EDA*) software helped to overcome this problem. Similarly, to address the hurdles highlighted above, a new software supported engineering automation approach is required as presented in Fig. 4.1. This section will present topics, where software can help with these tasks. Figure 4.9 summarizes all activities linked to a technology development to be supported by appropriate technology design methods and tools.

New ideas for process sequences are often based on or can benefit from previous developments. Knowledge management software can provide ways to access these previous development results in a structured manner. With such software, information can be retrieved faster, and previous results can be found and thus used more efficiently. Specialized software offers ways to view and search result data (e.g., materials, process steps, machines, and experiments) from different viewpoints, to categorize data under different aspects, and to provide ways to link data items together. Results and data that belong in the same context can be explored in the resulting network and valuable information can be gained.

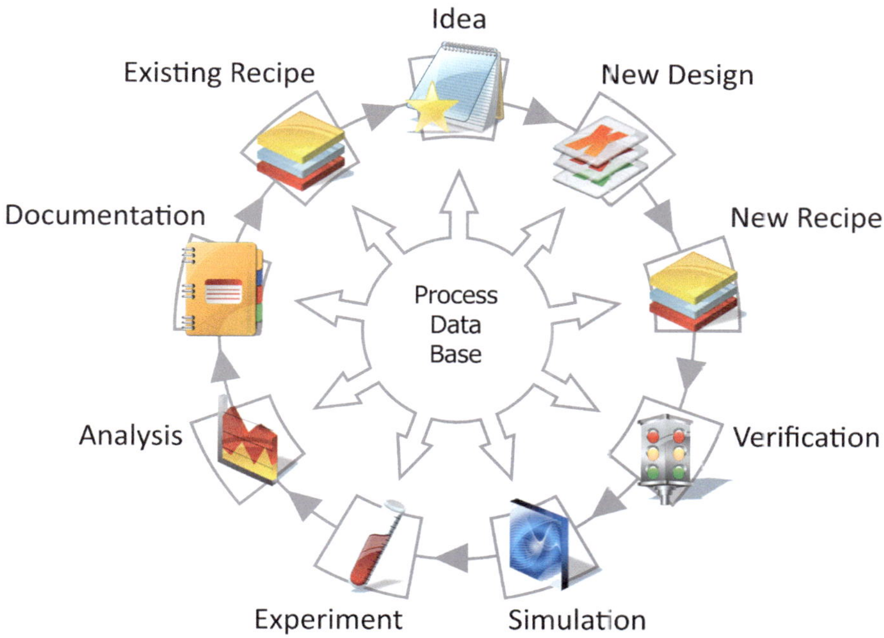

Fig. 4.9 Process development cycle

In the phase of assembling process steps to process sequences, software can ease the assembly, storage, and printing of new sequences. By providing access to previously assembled process sequences, a designer can use these as building blocks or modules in the newly developed sequence. The usage of standard building blocks can drastically reduce design time and mistakes in the design phase.

In the verification phase, a software system shows its real advantages. Most process design rules have a form that can be expressed in a machine-readable way in terms of rules (e.g., clean before deposition, and do not exceed 150°C as long as a polymer for lithography is deposited on the wafer). A domain expert can enter the rules for process steps to make them accessible for all engineers to check the newly developed process sequences. This means a supporting software system allows to manage rules, to connect rules with boolean terms (i.e., and, or, and not), and to check process sequences using these rules.

However, this check gives no indication of the functionality or the structure of the produced device. Technology Computer-Aided Design (TCAD) can provide at least an idea about the produced structure by simulating the process sequence. Most of these tools still need experts to write the simulation description files, so they are currently leveraged only by a small fraction of potential users. To support these "virtual fabrication" abilities to their full potential, a software system has to manage simulation models of different abstraction levels for process steps. It should provide support for multiple models for multiple simulators per process step. If a simple model is available for all basic process categories (e.g., deposition, etching, and lithography), the system is able to use the models for all process steps in the process sequence and generate an input file for the desired simulator. Currently simulation results are seen as stand-alone data. As an important improvement, a development system is able to manage the result files in combination with the process sequence. This enables the engineer to compare the expected results with the simulation results and with the real outcome. The knowledge gained from the comparison can then be used to improve the simulation model.

After verification, the device is produced in an experimental fabrication environment. In the first part of this phase, the process development system transfers the process sequence to the fab environment. This is done by simply printing out a runcard for the operators or by interfacing to the Manufacturing Execution System (*MES*) of the fab. On the other hand the system manages and documents last minute changes to the sequence such as parameter adjustments in the fab. Additionally lot splits and merges can be managed.

During and after processing, many measurements are done. These measurements often produce files such as pictures or simple texts containing rows and columns of data. A software system can manage these files, link related results together, and manage different versions of certain files (e.g., reports). Paired with flexible text and graphical retrieval and search methods, a development support system provides mechanisms to view and assess the accumulated data, information and knowledge from different perspectives. It provides insight into previous developments concerning information and time aspects.

Another important part of the data gathered during a development project is the informal knowledge. It consists mostly of the discussions between engineers about the experimental results. The engineering system channels and archives this knowledge because it is of immeasurable value to future projects.

A growing number of development activities in the industry are collaborative efforts. This increases the need to exchange information between partners or transfer process intellectual property (IP) from a vendor to a customer. Software systems can support this transfer while being selective to protect the intellectual property rights of the company.

In summary, process development software systems and the tools they consist of can assist with the execution of the development of new process sequences for new device architectures and devices. These tools support the whole engineering sequence—from the first device idea to the transfer of the resulting recipe into production or to a collaborating partner. Therefore, they close the loop and feed existing results into future ideas.

A new kind of software providing the functionality described in this section is delivered by Process Development Execution Systems (*PDES*). As introduced in, e.g., [9, 17], a PDES comprises a complete environment for all stages of process development—from initial concept to final experimental success and transfer into mass production. It provides a framework, which can be readily adapted to customer-specific situations and procedures. The purpose of a PDES is to manage and support the execution of development activities for high-tech manufacturing processes. A simplistic relational perspective is that a PDES is to the development of high-tech manufacturing processes what a Manufacturing Execution System (*MES*) is to the execution of volume production processes. MES are based on information technology systems that manage MEMS manufacturing operations in industrial production sites. MES comprises, e.g., the management of resources, the scheduling of manufacturing processes, the dispatching and the executing of manufacturing orders, and the collection and analysis of production data.

The significant difference between the two systems is that the emphasis of the PDES tool set is to deliver flexibility and experimentation means in a low volume environment, while the tools of a MES focus on less variance, tighter controls, and logistics in a high volume environment. The similarities in purpose of a PDES and MES are the traditional common-ground objectives of increasing traceability, productivity, and quality. With a PDES, the expectation and objective is to increase the quality of the developed manufacturing process, which is somewhat in contrast to the objective of improving production quality with a MES.

4.3.3 XperiDesk: A Process Development Execution System

The **XperiDesk** software suite (developed by the authors [16]) is a comprehensive application for the development of complex manufacturing processes in the semiconductor and thin-film MEMS device market. It supports process step and process sequence development with innovative tools. However, the power of XperiDesk is even more apparent in its tools that provide solutions for nontechnical challenges. A few examples are internal and external collaboration, documentation management and technology transfers, all of which ease time-to-market pressures by truly expediting the development process. As for subsequent developments, each can be progressively faster because XperiDesk provides complete reuse of data and documentation from previous developments. One of the innovative features of XperiDesk verifies the manufacturability of a recipe before actual manufacturing is carried out. This is achieved by using customizable rules to determine if the experiment could fail and/or damage equipment. An open interface to integrate external simulation or calculation software enables efficient structural and other assessments like costs of designed devices. Robust tracking mechanisms support the management of all instances of experimental and verification data as well as the execution of iterative improvement steps. Furthermore importing configurations can be created and scheduled to pull data from diverse sources into a holistic and comprehensive view. XperiDesk additionally makes the degree of distributed collaboration transparent and secure. A user/role security model, specifically designed for IPR protection on a single item level, manages the distinct demarcations of various collaborations. Another feature provides selective, automated, data exchange to collaborating

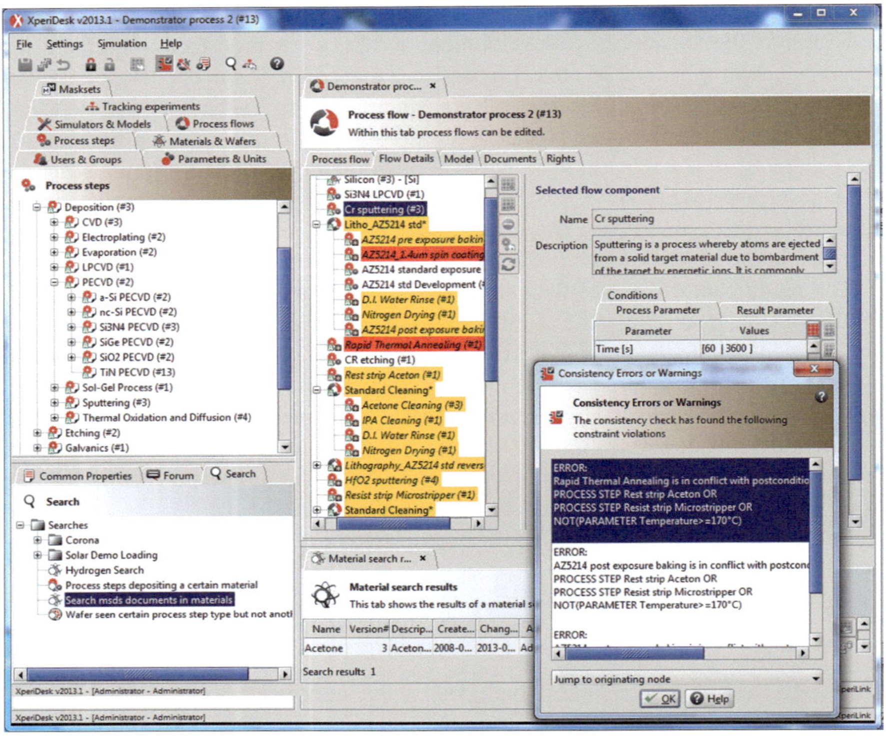

Fig. 4.10 XperiDesk process flow editor with highlighted manufacturability issues

partners and customers while preserving the IP protection. The whole software suite consists of different modules, each focusing on one class of challenges:

XperiCipe enables the reuse of recipes and documentation from previous developments efficiently. It supports the management of data, information and knowledge during the initial concept stages as well as life cycle aware maintenance of manufacturing flows. Data is managed from the minuscule conversion of values into different units to the organization of entire manufacturing flows. Versioning and life cycle management ensures the ability to change recipes at any time without damaging historical process documentation or production. Design entities can be arranged in multiple hierarchies and categories. Sophisticated inheritance mechanisms permit the propagation of design properties and data for each entity throughout the entire hierarchy structure. Figure 4.10 presents a view of the process flow editor with manufacturability issues highlighted. The power of this module increases with its use due to the support for the reuse of already developed design elements.

XperiFication provides a virtual assessment of newly created or modified production recipes. It assesses the manufacturability of a recipe via a consistency

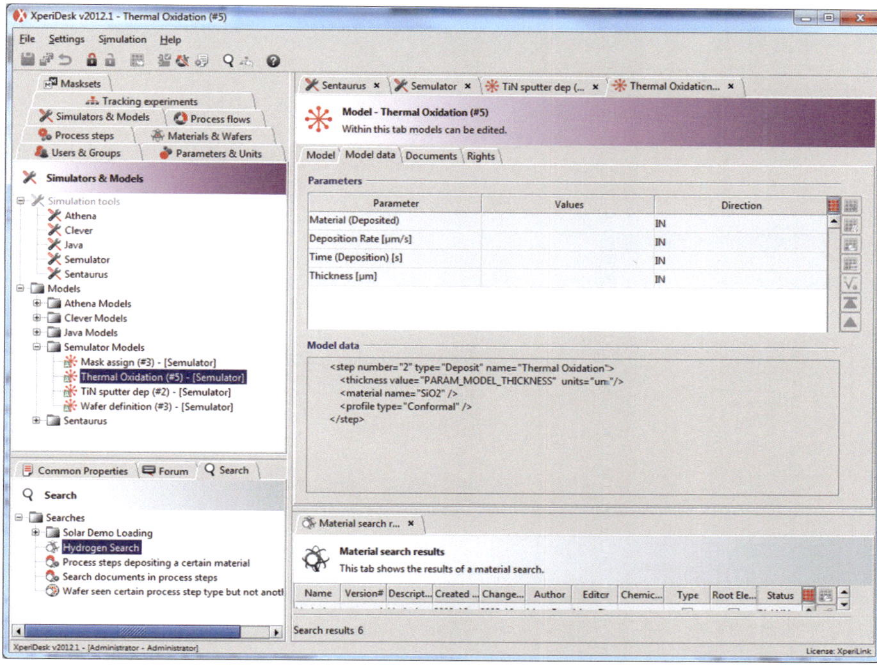

Fig. 4.11 XperiSim screenshot

check engine that uses customizable rules to evaluate the compatibility requirements of process steps. Figure 4.10 presents an example of such a check.

XperiSim's generic simulation and calculation interface achieves process calculations (TCAD simulations, cost assessments, etc.). It manages models for multiple simulators and calculators that can be adapted by experts for company wide use. That allows engineers without prior experience to run all kinds of flow-based calculations (device structure generation, cost, etc.). XperiSim builds the necessary input without requiring any programming or intimate calculator engine knowledge. Figure 4.11 shows an interface to Coventors's Semulator [3]. XperiSim archives result files in the underlying document management system, and the result files are accessible from any client, since they are managed in the context of the sequence.

XperiLink supports the tracking of experimental verification results. It automatically collects files, loads data from the file system, MS Excel, or other databases, and provides navigation and retrieval of data via multiple criteria searches, filters, and views. Therefore it breaks down the typical data silos and provides a consistent, comprehensive view of all data created during experimental verification including the timeline. The data can have multiple levels of detail to manage parameters, etc., with the added benefit of viewing results from multiple perspectives.

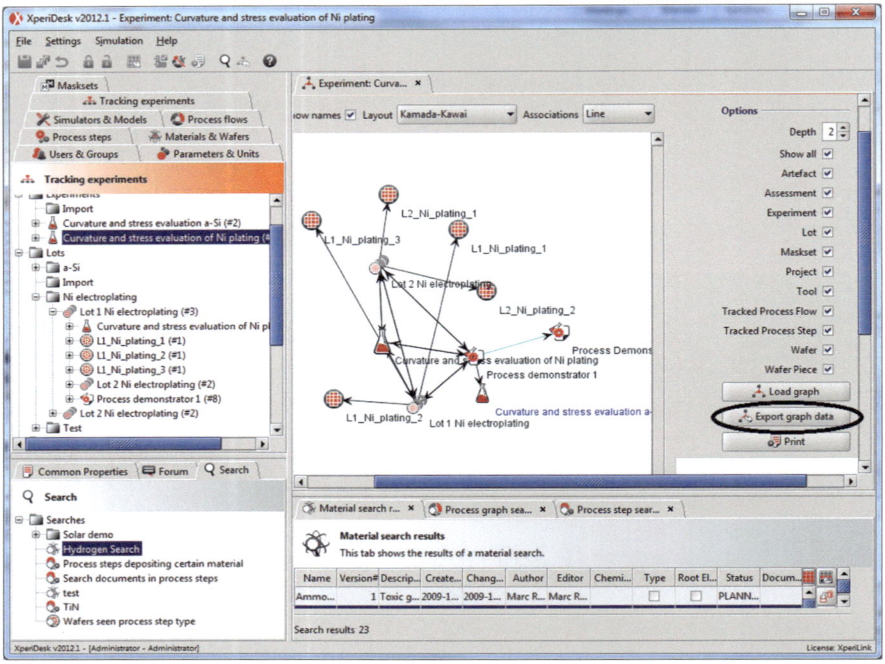

Fig. 4.12 XperiShare screenshot

XperiShare provides the selective, automated exchange of development data between different partners. Selective export and import functionality allows the bundling of IP packages and the transfer of various simulation and experimental verification results. XperiShare increases the collaboration efficiency to achieve an expedited fabrication ramp-up. Figure 4.12 shows a screenshot of the system.

4.4 Quality Considerations

Due to the increasing requirements for MEMS products innovation-driven companies are forced to achieve a higher level of quality combined with minimal cost and shortened production cycles. To assure product quality in all phases of the innovation process an appropriate quality management (QM) method has to be used. A quality management system covers all quality procedures and quality assurance activities.

In this context the Design for Six Sigma (DfSS) method turned out to be promising. An early focus on the principles of DfSS may additionally result in a significant reduction in overall product cost. DfSS applied already in the early stages of design defines the quality as an essential design goal. In this section only

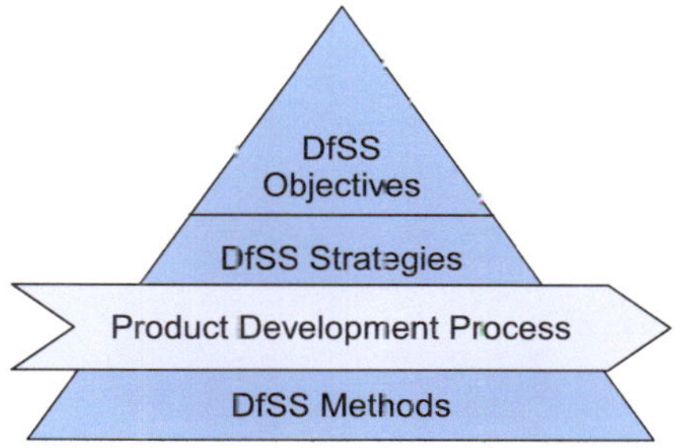

Fig. 4.13 Design for Six Sigma

a few general considerations regarding DfSS will be made since the complete issue is subject to several detailed book publications, see, e.g., [5, 23].

4.4.1 *Design for Six Sigma*

Design for Six Sigma (DfSS) is a supportive approach which focuses on the optimum planning and control of the engineering of new products or processes. The term Six Sigma refers to the standard deviation of a process mean value and stands for a close to zero-defect rate of more than 99.99966 %. The DfSS concept interwoven into the product development flow is shown in Fig. 4.13. The top level describes the DfSS objectives. Product indicators, target values, and procedures can be derived from these objectives. Among the main objectives is the requirement for error free, robust, and reliable products as well as competitive advantage through consistent customer orientation and manufacturing with high resource efficiency (also called frontloading). In [5] the individual targets are explained in detail.

DfSS strategy level describes the principles and strategies to achieve the objectives. The primary goal is the elimination of errors in early stages of development and thus the achievement of a zero-defect level. Reference [5] addresses this issue in detail. The strategy level also describes phases or steps for DfSS development processes. Acronyms denote the most popular strategies. Frequently proposed DFSS phases are, e.g., IDOV (Identify, Define, Optimize, Validate), or DMADV (Define, Measure, Analyze, Design, Verify). The selection of the proper process steps depends inter alia on whether it is a new design or redesign project. In the context of MEMS product engineering an approach according to DICOV (Define, Identify, Characterize, Optimize, Validate) as suggested in Fig. 4.14 is very common.

The DfSS process level represents the application scope of DfSS, i.e., the phases of a product engineering process where DfSS is applied to. Finally, the whole concept is supported by the DfSS method level containing a number of methods that are used within to implement the goals and strategies. The objective of customer orientation is picked up by using the *Voice of the Customer (VOC)* method. Subsequently the method *Quality Function Deployment (QFD)* is used to

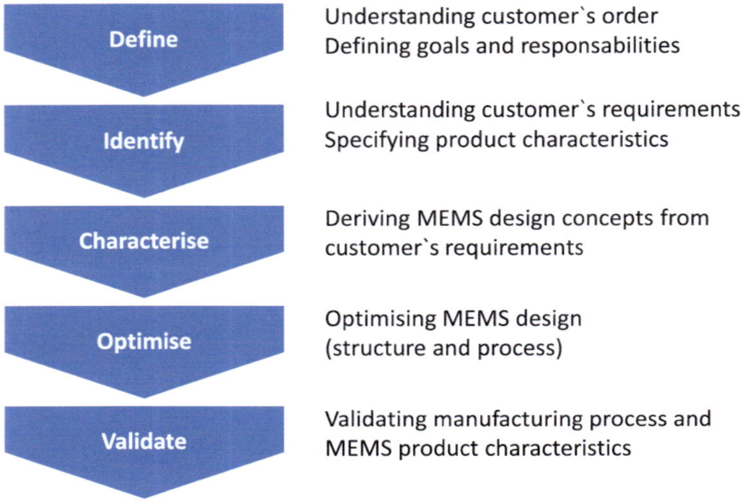

Fig. 4.14 DICOV stategy

transform customer's requirements into a technical product development process with engineering characteristics.

VOC is the starting point for the DfSS goal *"Competitive Advantage through Customer Orientation,"* as well as for the subsequent Quality Function Deployment Method. With VOC a systematic collection and structuring of the customer's requirements at the beginning of a development can be reached [5]. The subsequent application of QFD is to ensure that not only the view of the manufacturer, but on first hand also the needs of customers, users and of the market are understood and will be implemented in the product. Initially the requirements of customers given in VOC are systematically transferred in technical requirements for products. Finally the technical specs are translated into process and production requirements. For this procedure QFD provides a consistent model of matrices, called"*Houses of Quality"* see Fig. 4.15.

The matrices are divided into several "rooms," each of them representing the specific issues and perspectives. The focus of the considerations in *House of Quality 1* are the customer requirements identified, for example, from VOC and their transformation into technical MEMS product features. Based on the required product properties possible MEMS design flows will be developed and evaluated. The product features of this stage serve as input for the *House of Quality 2*. At this level component properties will be derived, which are passed as input variables for the manufacturing planning in the *House of Quality 3*. The determination of manufacturing process steps and parameter settings to meet the quality requirements of components and subcomponents is the central task of this stage. Finally in the *House of Quality 4* describes product validation and test in detail. The interaction of the four stages represented by the *Houses of Quality* is shown in Fig. 4.16.

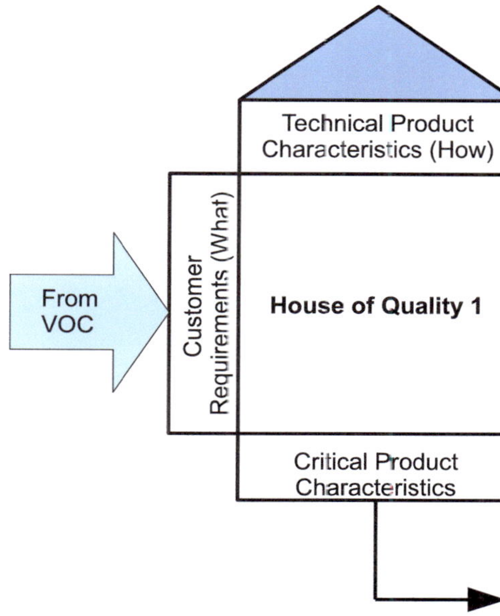

Fig. 4.15 DfSS: house of quality

QFD methods can be used during the entire engineering process. At the beginning the QFD process can be applied in a top-down approach and at the end with a detailed specification of MEMS components and process steps a bottom-up method will be appropriate. Depending on the phase of the engineering process QFD is performed with varying degrees of detail. The respective phases can be carried out recursively and thus earlier decisions can be clarified or corrected. Consistently applied and implemented QFD is a method by which a MEMS product can be developed quickly, systematically, and can concurrently be improved in a goal-oriented manner. The strength of this procedure lies in particular in the constant customer focus.

Figure 4.17 shows the interaction of QFD with other methods. After VOC is performed and passed as input to QFD, certain results can be checked with simulation, DoE (Design of Experiments) or can be reviewed and clarified with analytical modeling. Functions that had been developed with QFD can be validated with risk analysis methods like DRBFM (Design Review Based on Failure Mode) or FMEA (Failure Modes and Effect Analysis). The results of QFD in the Houses of Quality 3 and 4 eventually can be passed to, e.g., SPC (Statistical Process Control). Exact explanations for each of the methods can be found in [5].

The following objectives should be achieved with DfSS:

- a more or less complete understanding of the (main) product requirements
- an early identification of challenging areas at the beginning of the design
- a clear determination of the critical design parameters (CDPs)
- a flow-down of the product requirements and flow-up of the product capabilities

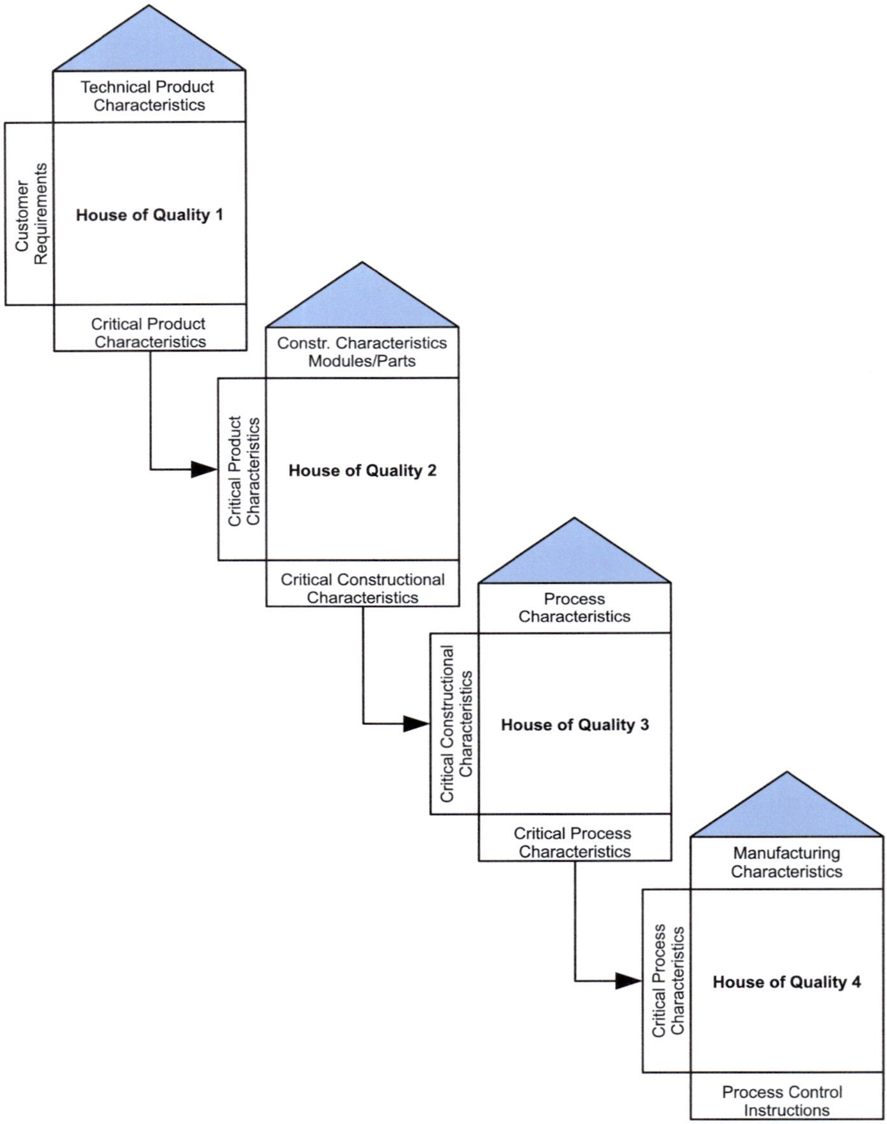

Fig. 4.16 DfSS: four houses of quality

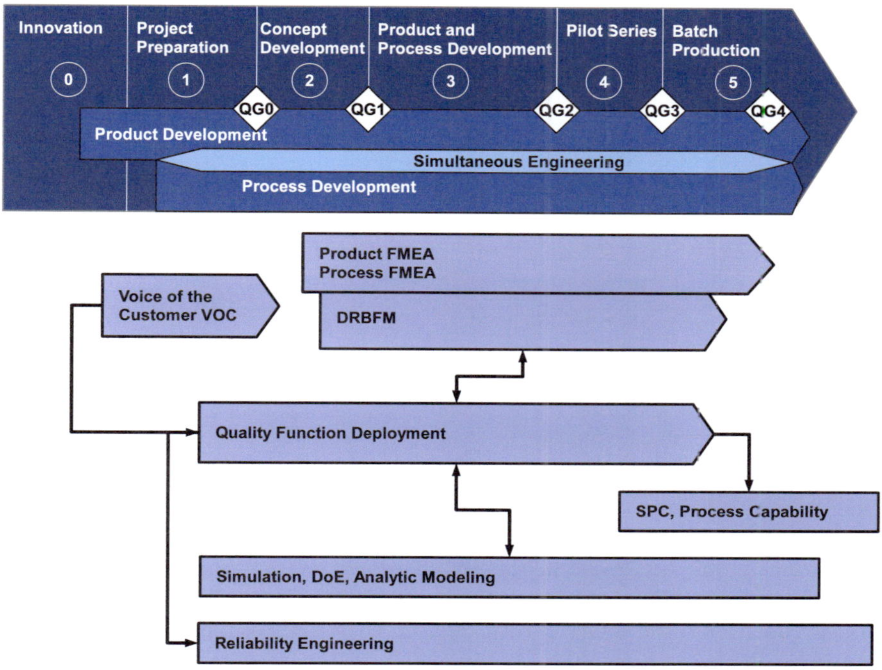

Fig. 4.17 Design for Six Sigma and product engineering [5]

4.5 Customer-Oriented Product Engineering

4.5.1 The Customer in the MEMS Industry

Before outlining the influence of the customer on the Product Engineering process it is necessary to define what is actually meant by the term customer in this context. Depending on the position in the MEMS product engineering chain the term "customer" fits on many very different parties. The customer in terms of the new PE methodology is the party that defines the technical requirements of the product. For MEMS products this is clearly not the end user (e.g., the driver of a car who is only interested in user interfaces and not in technical constraints) nor the OEM (who is mainly interested in the functionality and not in the operating principles or technologies used to achieve this functionality). The customer in a MEMS context can be rather found on the level of the 1st tier supplier or on a product definition group that acts as a mediator between the requirements given from OEM and end user and the technical specifics coming from the fabrication facilities at hand. Figure 4.18 gives an impression where this kind of customer is located.

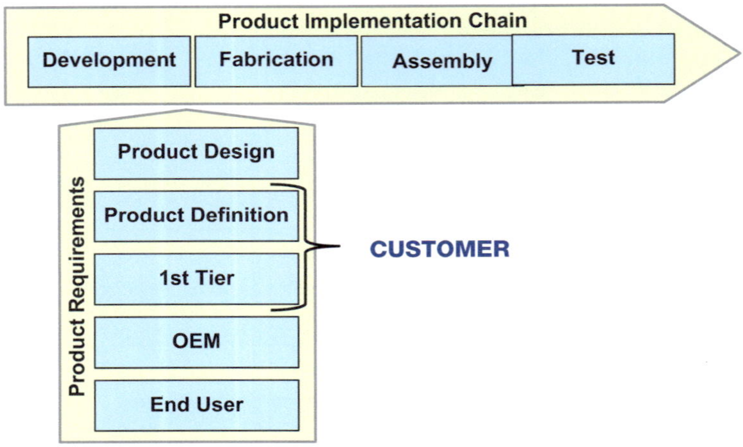

Fig. 4.18 Customer location in the MEMS PE chain

Customers on this level are highly interested in the technical aspects and constraints of the MEMS product as they are usually in charge of the whole process from product definition to product distribution. Therefore this type of customer is the only entity in the chain capable of keeping track of all the complex requirements and constraints throughout the whole PE process.

4.5.2 Customer Involvement

The customer—as described in the previous section and published in, e.g., [8]—is in charge of the Product Engineering process. The customer is the party that defines the product by specifying the functionality, as well as the requirements and constraints of the intended business case. Apart from technical aspects the customer is also the initiator and manager of the whole Product Engineering process. This involvement is outlined in Fig. 4.19.

Product realization can be divided into two phases. The first phase where the specifications of the product will be turned into a detailed set of fabrication instructions fulfilling all requirements and constraints. The second is the fabrication and assembly phase where the fabrication instructions are executed to generate an arbitrary number of copies of the intended product, each guaranteed to meet all requirements and constraints. The customer takes the central role in this scenario. It is typical for MEMS that the roles of the parties involved may differ between projects. As the only one with in-depth knowledge of the intended product, the customer will supervise and guide the whole product engineering cycle. For an automotive electronics product the specific role of the customer arises in MEMS because constraints from the application domain are usually very complex and only

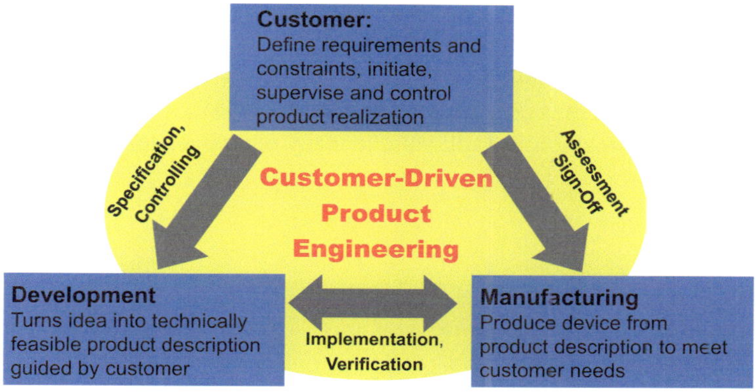

Fig. 4.19 Role of the customer in MEMS product engineering

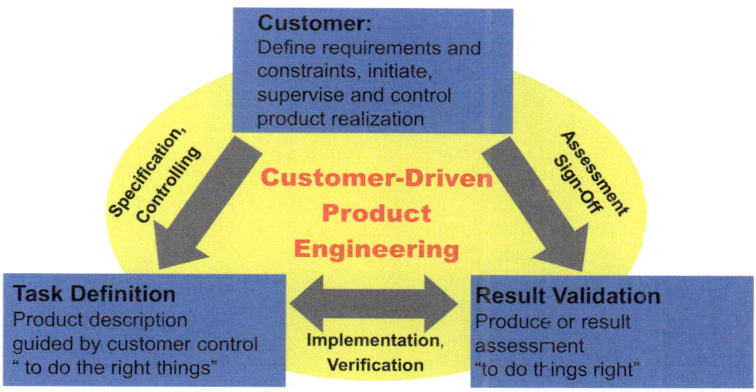

Fig. 4.20 Customer involvement in the development phase

fully understood by the customer. On the other hand the inherent complexity and diversity of the manufacturing technology make it a complicated task to select and to configure an appropriate manufacturing process. The development substep outlined in Fig. 4.19 can itself be split into a similar interaction triangle presented in Fig. 4.20.

Efficient and non ambiguous communication between the customer and all other parties is therefore of utmost importance for shortening the Time to Market. In more classical "offline" product innovation (e.g., the white goods sector), the customer just specifies a product and more or less randomly selects a possible technology supplier. From then on design and fabrication are performed without customer interaction. This simple scenario is usually not applicable to the MEMS sector. In a customer-oriented approach the product specification is gradually turned into a set of manufacturing instructions by direct involvement of the customer. The customer actively takes part in all development activities to ensure that the solution will fulfill the specific requirements of the application, whether explicitly specified or not. This is extremely important for MEMS products, because even small deviations might directly affect economical aspects and quality. Therefore PE can only be performed in close partnership between customer and supplier, rather than in a traditional customer / supplier relationship still valid in many other industrial sectors.

References

1. Anderson, R.E.: Phased product development: friend or foe? Business Horizons **39**(6), 30–36 (1996)
2. Barkai, J.: Accelerating Science-led Innovation for Competitive Advantage. IDC Manufacturing Insights, Framingham (2012). http://www.winvale.com/pdfs/white_paper_module/IDC_Accelerating_Science-led_Innovation.pdf. Accessed 21 Aug 2013
3. Coventor, Inc.: Coventor home page. http://www.coventor.com/ (2013). Accessed 10 April 2013
4. Gajski, D.D., Thomas, D.E.: Introduction to silicon compilation. In: Gajski, D.D. (ed.) Silicon Compilation, pp. 1–48. Addison-Wesley, Reading (1988)
5. Gamweger, J., Jöbstl, O., Strohrmann, M., Suchowerskyj, W.: Design for Six Sigma. Carl Hanser Verlag, München (2009)
6. Hahn, K.: Methoden und Werkzeuge zur fertigungsnahen Entwurfsverifikation in der Mikrotechnik. VDI Verlag, Düsseldorf (1998)
7. Hahn, K., Hedrich, L., Salfelder, F., Kremer, H., Völklein, F., Korb, W.: Mides - Mikrosystemtechnik Design Flow für fr KMU. In: Edaworkshop 2012 Hannover (2012). EdaWorkshop 2012. VDE Verlag GmbH, Berlin (2012)
8. Hahn, K., Schmidt, T., Mielke, M., Brück, R., Ortloff, D.: Micro and nano product engineering using data management for silicon-based fabrication process development. In: 9th IEEE Conference on Nanotechnology, pp. 337–340. IEEE-NANO 2009. IEEExplore, New York (2009)
9. Hahn, K., Schmidt, T., Mielke, M., Brück, R., Ortloff, D., Popp, J.: Distributed product engineering for microsystems devices. In: Proceedings of the 17th International Conference on Mixed Design of Integrated Circuits and Systems (MIXDES), pp. 453–457. IEEExplore, New York (2010)
10. Hahn, K., Wagener, A., Popp, J., Brück, R.: Process management and design for MEMS and microelectronics technologies. In: Proceedings of SPIE: International Symposium on Microelectronics, MEMS, and Nanotechnology 2003, Perth (2003). SPIE Perth 2003
11. Inc., S.: Jmp statistical discovery. http://www.jmp.com/ (2013). Accessed 10 March 2013
12. Jiju, A.: Design of Experiments, 1st edn. ButterworthHeinemann, Oxford (2003)
13. Kahng, A.B., Lienig, J., Markov, I.L., Hu, J.: VLSI Physical Design: from Graph Partitioning to Timing Closure. Springer, Dordrecht, Heidelberg, London, New York (2011)
14. Mead, C., Conway, L.: Introduction to VLSI Systems. Addison-Wesley, Reading (1980)
15. Minitab Inc.: Minitab statistical software. http://www.minitab.com/products/minitab/default.aspx (2005). Accessed 10 March 2013
16. Ortloff, D., Popp, J., Wagener, A., Hahn, K.: Breaking through the process development barriers. In: Proceedings of the 12th International Conference on the Commercialization of Micro and Nano Systems, Puerto Vallata (2008). COMS 2008
17. Popp, J., Ortloff, D., Schmidt, T., Hahn, K., Mielke, M., Bruck, R.: Sampling process information from unstructured data. In: Advanced Semiconductor Manufacturing Conference (ASMC), 2011 22nd Annual IEEE/SEMI, pp. 1–6. IEEExplore, New York (2011). doi:10.1109/ASMC.2011.5898184

18. ReVelle, J.B.: Quality Essentials – A Reference Guide from A to Z, 1st edn. ASQ Quality Press, Milwaukee (2004)
19. Schmidt, T., Ortloff, D., Hahn, K., Wagener, A., Popp, J., Brück, R.: Management and verification for MEMS fabrication processes. In: Proceedings of SPIE: Micromachining and Microfabrication Process Technology X, vol. 6109, San Jose (2006). Photonics Europe 2006
20. Sequin, C.: Managing vlsi complexity: an outlook. Proc. IEEE **71**(1), 149–166 (1983)
21. Wagener, A., Popp, J., Schmidt, T., Hahn, K., Brück, R., Ortloff, D.: Environment for design and verification of mems fabrication processes. In: Proceedings of the MST 2005. VDE Verlag GmbH, Berlin (2005)
22. Wagener, A., Schmidt, T., Popp, J., Hahn, K., Brück, R., Ortloff, D.: Process design and tracking support for MEMS. In: Proceedings of SPIE: Micromachining and Microfabrication Process Technology X, vol. 6109, San Jose (2006). Photonics West 2006
23. Yang, K., Haik, E.: Design for Six Sigma. McGraw-Hill, Columbus, OH (2003)
24. Yole Développement: MEMS Foundries 2009. Yole Développement, Lyon-Villeurbanne (2009)
25. Yole Développement: Status of the MEMS Industry. Yole Développement, Lyon-Villeurbanne (2010)

Chapter 5
Methodology

Abstract This chapter gives an overview of the method composition and then dives deep down into the details of the individual methodology blocks. For all Stages and Gates a detailed description of their purpose, their constraints, their inputs, and their outputs is given. Example checklists of deliverables and evaluation criteria are shown. Furthermore flowcharts that lead through the blueprint business processes are presented in combination with tools to execute them. This chapter provides only a blueprint for custom innovation processes. For a sample implementation of the methodology, see Part III of this book.

5.1 Overview

The assessment of baseline methods in Sects. 2.3.1.3 and 2.3.2.4 revealed that a combination of the product development method Stage-Gate® and the project management method PRINCE2® is a promising approach for Product Engineering in the context of MNT. Both methods treat their efforts (product development and project management) as processes, emphasize on quality aspects, and are used by a widespread commercial and governmental community. Additionally the methodologies are deliverable oriented rather than task oriented. Furthermore the product development method Stage-Gate® is more or less the standard in the organizations which participated in the market study performed by the CORONA project [10]. The efforts of the MEMS Industry Group and its members to define a Technology Development Process Template [7] resulted in similar findings. The T^2M methodology introduced in Sect. 2.3.3 is used as a tool for detailed development planning and coordination of development efforts between multiple independent partners. T^2M is based on a notion of tasks where each task produces or modifies a deliverable. Therefore the binding element of all three methodology components are the Deliverables.

Stage-Gate® focuses on product development tasks spanning the whole range from the first idea to the launch of the product into the market. It defines the

D. Ortloff et al., *MEMS Product Engineering*, DOI 10.1007/978-3-7091-0706-5_5,

deliverables as well as the evaluation criteria to assess the viability of the initial product idea. Therefore it deals with the question *What* to develop and provides guidance on *How* to proceed. The main characteristics of Stage-Gate® are that it divides the project into several consecutive stages, where the actual development work is done, and gates, where this work is assessed and go/hold/kill decisions are made whether to continue investing in the project or not.

PRINCE2®, on the other hand, is a project management method with a strong focus on efficiency and quality aspects. It provides a framework to perform the various project management processes within the Product Engineering methodology. It supplements the Stage-Gate® method nicely with detailed guidance on the *How*.

The combination of the Stage-Gate® and PRINCE2® methodologies has been previously proposed in [9] and for reference and as an overview explanation is outlined in Fig. 5.1. The approach proposed in [9] is tailored for fabless MEMS IP companies and has five stages where efforts and resource requirements increase with each subsequent stage. This method can be viewed as one concrete incarnation for this special business environment of the more general approach proposed by this book.

The method proposed in [9] follows a pattern that is also picked up within this book: After an initial assessment of a product idea in gate 1 the first stage analyzes the market potential and the general technical feasibility. The decision whether or not to continue with product development is based on the assessment performed in gate 2. Stage 2 starts and initializes a development project by appointing the project responsible persons, setting up the project infrastructure, setting up an initial business case, and developing the different plans (e.g., quality plan, project plan, stage plan) according to PRINCE2® . This stage is followed by gate 3, only checking whether the boundary constraints determined during gate 2 are still valid. If this is the case, it leads to the stage 3 in which the product idea is evaluated in more detailed technical, market, legal, and financial investigations. This includes building a more solid business case and a technical concept prototype (approx. 10 % development effort). Gate 4 is also called "money gate," as stage 4 requires huge investments into the project. This stage is where prototype development and initial lab testing takes place. Result of stage 4 is a production grade prototype of the product. Several intermediate quality assessment points (e.g., quality audits) are also part of this stage. Gate 5 reviews the completeness and the quality of the prototype development. The final development stage continues with intensive testing, broadening the process windows, and wrapping up the other product components. Finally during the gate 6 assessment the decision to introduce to the market and roll-out the product is taken. The whole progression of these efforts is managed by the PRINCE2® management processes and all Gates are potential redo/hold/stop points. Fewer stages would add a significant risk because each stage would become bigger requiring more commitment to that stage.

This specific process for MEMS process IP companies can be generalized into the generic approach defined and implemented here. The most crucial extension of this approach is the integration of T^2M (described in more detail in Sect. 2.3.3)

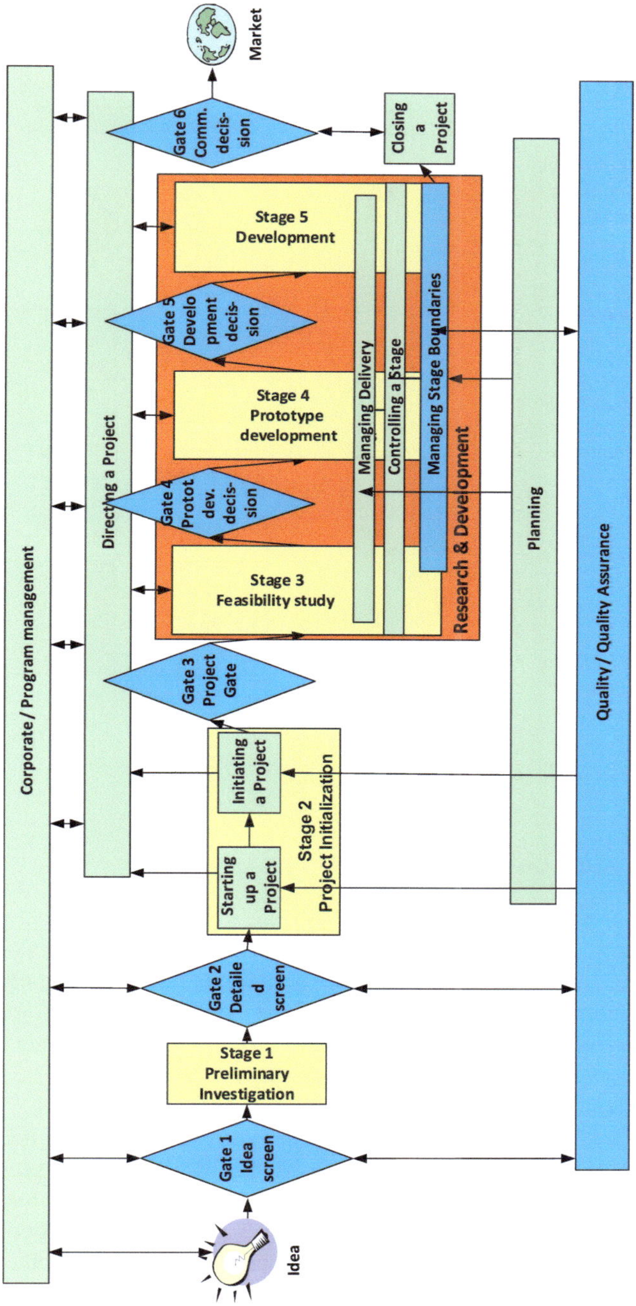

Fig. 5.1 Sample Stage Gate® Methodology assisted by PRINCE2® processes as proposed in [9]

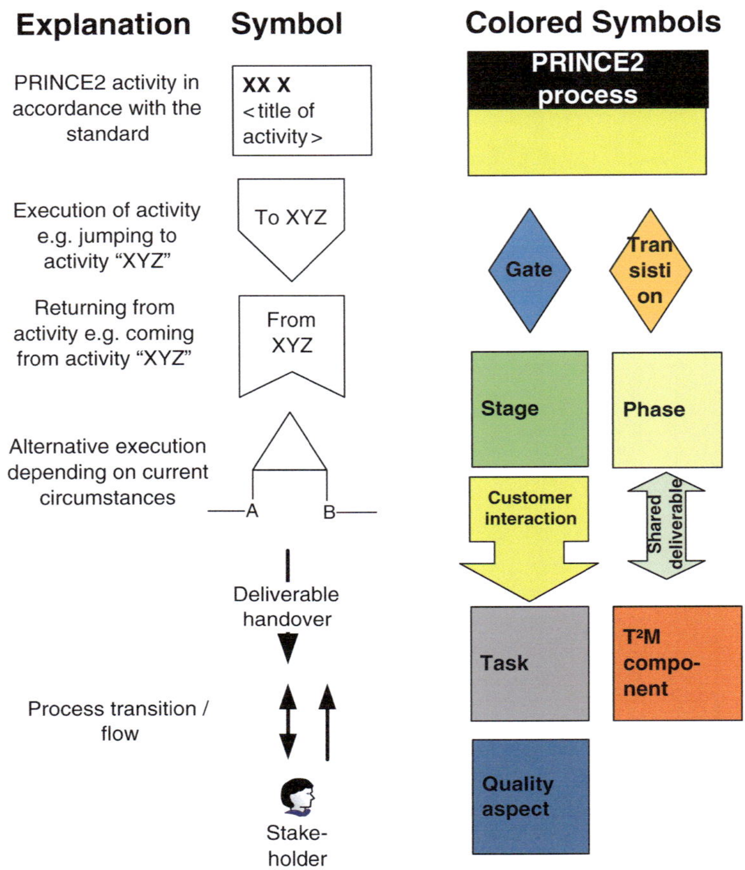

Fig. 5.2 Symbols used in process drawings

to add more flexibility by being recursively applicable and being more flexible and more open for customer interactions. The requirement for these more tightly coupled customer interactions is introduced in, e.g., [5, 6]. This more general approach is described in the following sections. These descriptions and drawings are using the symbols and colors summarized in Fig. 5.2. Deviations from these will be explicitly mentioned or are indicated within the drawings themselves.

Both PRINCE2® and Stage-Gate® use many different terms and items to the development process. Most of them need to be specifically tailored according to the application area of the project. Additionally, the combination of three different methods (PRINCE2®, Stage-Gate®, and T^{2}M) with possibly conflicting terms requires a clear definition of the actual meaning of the terms in the context of this book. The details of the definitions of terms can be found in the Glossary at the

Table 5.1 Terms and definitions overview

CORONA	PRINCE2®	Stage-Gate®	T^2M
Deliverable	Product	Deliverable	Deliverable
Stage	Stage	Stage	–
Gate	–	Gate	–
Transition	–	–	Transition
Task	–	–	Task

end of this book. The following table gives an overview of the partly overlapping definitions (Table 5.1).

5.2 Method Composition

Section 5.1 gave a brief overview of a sample Product Engineering flow using a combined Stage-Gate® / PRINCE2® approach. As the MNT industry is very diverse, there are many different scenarios for Product Engineering. For very small projects with limited budget and limited time the sample Product Engineering flow might cause too much overhead. On the other hand there are also large innovation projects with many partners that will run for several years. In this case the sample process might not be sufficiently detailed and not divide into enough steps and requires additional stages and gates. For this reason the approach has been modularized, so that for each project an individual Product Engineering flow can be tailored. The modular flow is divided into four phases each containing two Stages. Depending on the project at hand the two Stages of each phase can be merged into a single stage. Consequently this allows defining Product Engineering flows with a minimum of four stages up to a maximum of eight stages. Figure 5.3 shows the minimum flow with the merged stages (in the upper flow) and the maximum flow with all stages expanded (in the meandering flow). The purpose and the deliverables of each stage are explained in the following sections.

While the merged approach in the overall innovation process is guided by the Stage-Gate® process, all stages and gates starting from the StartUp phase are executed by the PRINCE2® (sub-) processes according to Stage-Gate® practices. By definition projects managed with PRINCE2® are organized in stages and product-based planning is focused on the deliverables of each stage. The method defines mechanisms how to handle deviations and escalation situations and offers several tools and techniques to guide the ventures—perfectly complementing the staged development approach of the Stage-Gate® method. This is explained in the next section. Inside the stages the execution of the PRINCE2® work packages identified by the work breakdown structure (WBS) is guided by T^2M modules. Therefore the processes of the "Managing Product Delivery" group use the techniques of T^2M and especially the planning process performs the planning setting up the appropriate T^2M modules. Via the T^2M the customer involvement and cooperative engineering

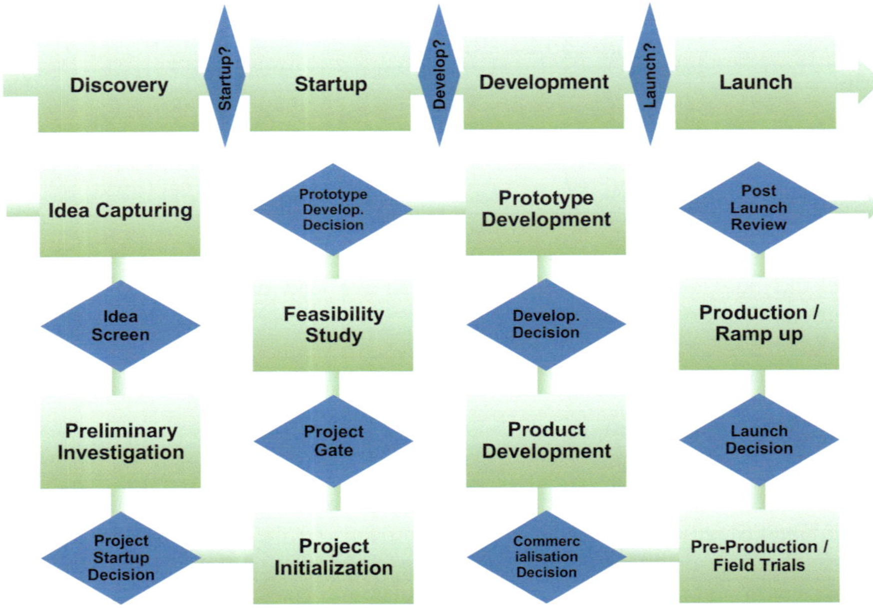

Fig. 5.3 Complete development method

is managed. The modules are used to implement best practices and have to be tailored individually to the project and the involved partners. Sample modules will be explained within the individual stage descriptions.

All processes, controls, and techniques of the PRINCE2® standard are integrated into the new method and all management processes are executed accordingly. Therefore the following descriptions focus on the steps, the integration of the methods, and the development tasks. The details of the PRINCE2® management processes are only given as far as there are deviations from the standard. For details about the management efforts the literature [1, 8] can be consulted and an overview in the form of a process map is given in Fig. 8.1. Parts of this figure will be extracted later on for better readability.

Figure 5.4 gives a general overview of the method and where the PRINCE2® processes are utilized. During the first part of the startup phase the processes "Starting up a Project" and "Initializing a Project" are executed. From then on all further Stages and Gates are controlled by the appropriate processes (e.g., "Controlling a Stage") until the project is finally closed using the "Closing a Project" process.

The fully integrated method with a medium level of detail is presented in Fig. 5.5 (left part of the process), Fig. 5.6 (middle part of the process), and Fig. 5.7 (right part of the process). The figures are supposed to present the general approach integration and highlight the overall flow. They are included to present an overview rather than showing readable details. Readable extractions from these figures are used in the

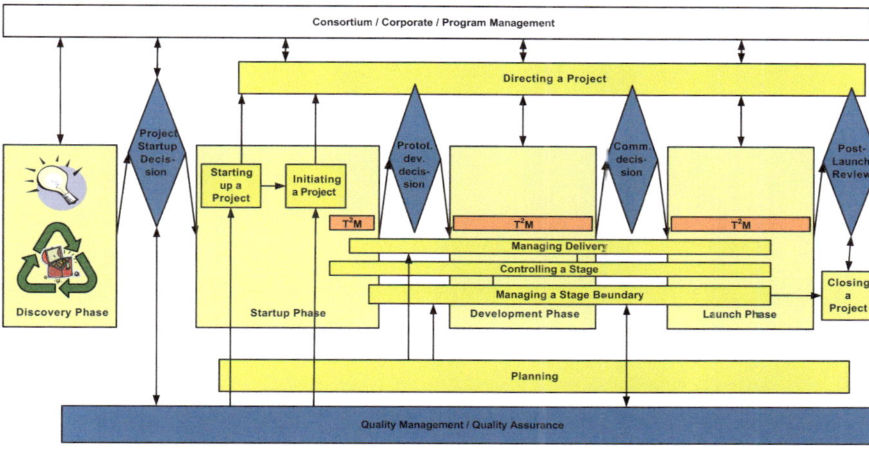

Fig. 5.4 Overview of PRINCE2® integration with T^2M and Stage-Gate®

following sections to explain the process in more detail and focus on the boundaries of the phases of the process.

5.3 Discovery Phase

The first step towards a new product is the development of a product idea, vision, or concept. If there is no product idea, there is nowhere to start and even if there is an idea it might still be a bad one. The purpose of the *Discovery Phase* lies in capturing a product idea and evaluating its feasibility.

These two tasks are reflected by two stages: *Idea Capturing* and *Preliminary Investigation*. As outlined in the methodology basics, it is possible to merge the two stages of every phase into a single stage. In this case this would mean omitting the intermediate *Idea Screen Gate*. Depending on the environment this might make sense, for instance in a highly innovative research context where product ideas are generated in parallel with research activities, like it is the case in academic settings. For almost all other settings it is, however, advisable to retain the two-stage approach.

The *Discovery Phase* is done almost exclusively by the customer company. For the *Preliminary Investigation Stage* potential development partners might be included for consulting.

5.3.1 Idea Capturing Stage

The first step in product engineering is the capturing of the product idea. Naturally, this is a stage that is difficult to control and does not really fit into a Stage-Gate®

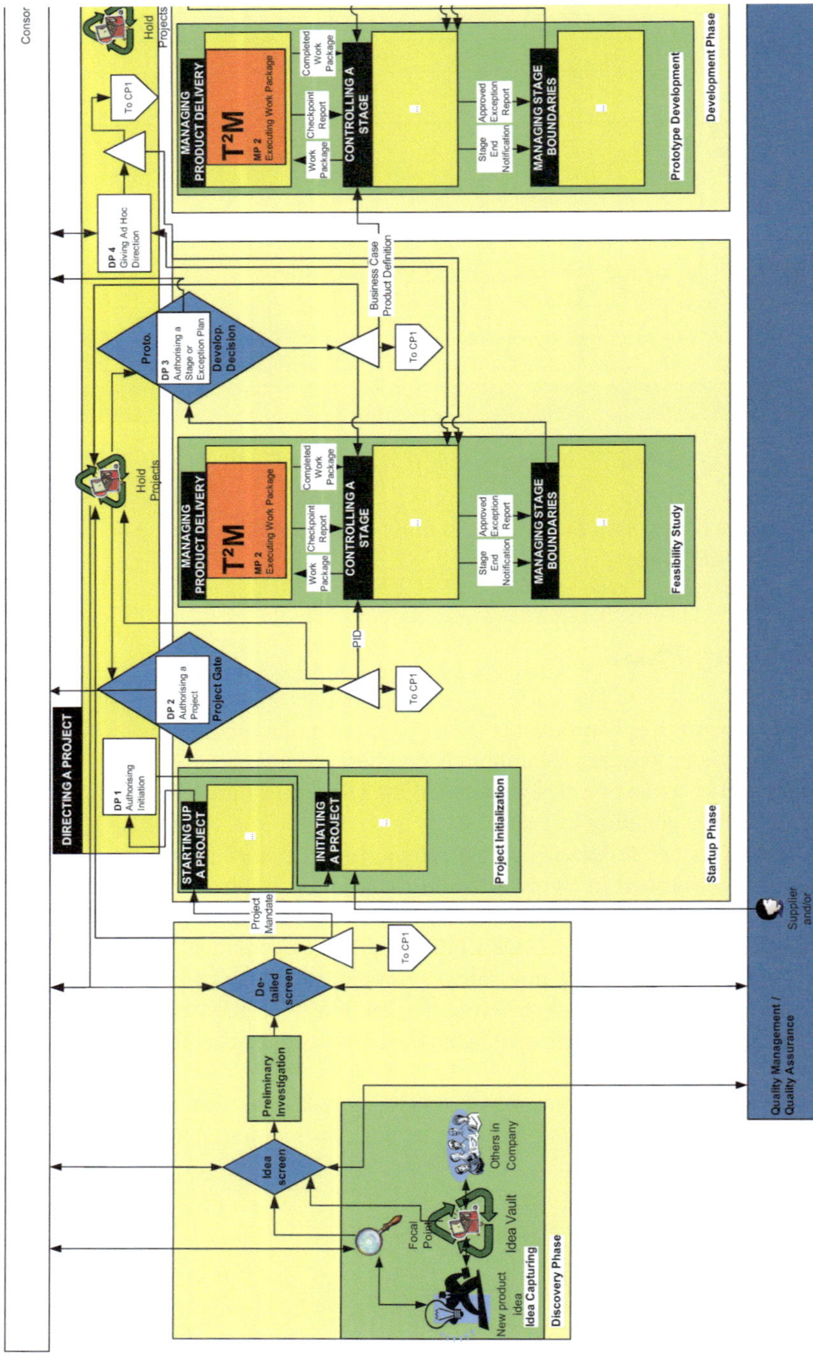

Fig. 5.5 Product engineering method with medium detail *left* part for overview

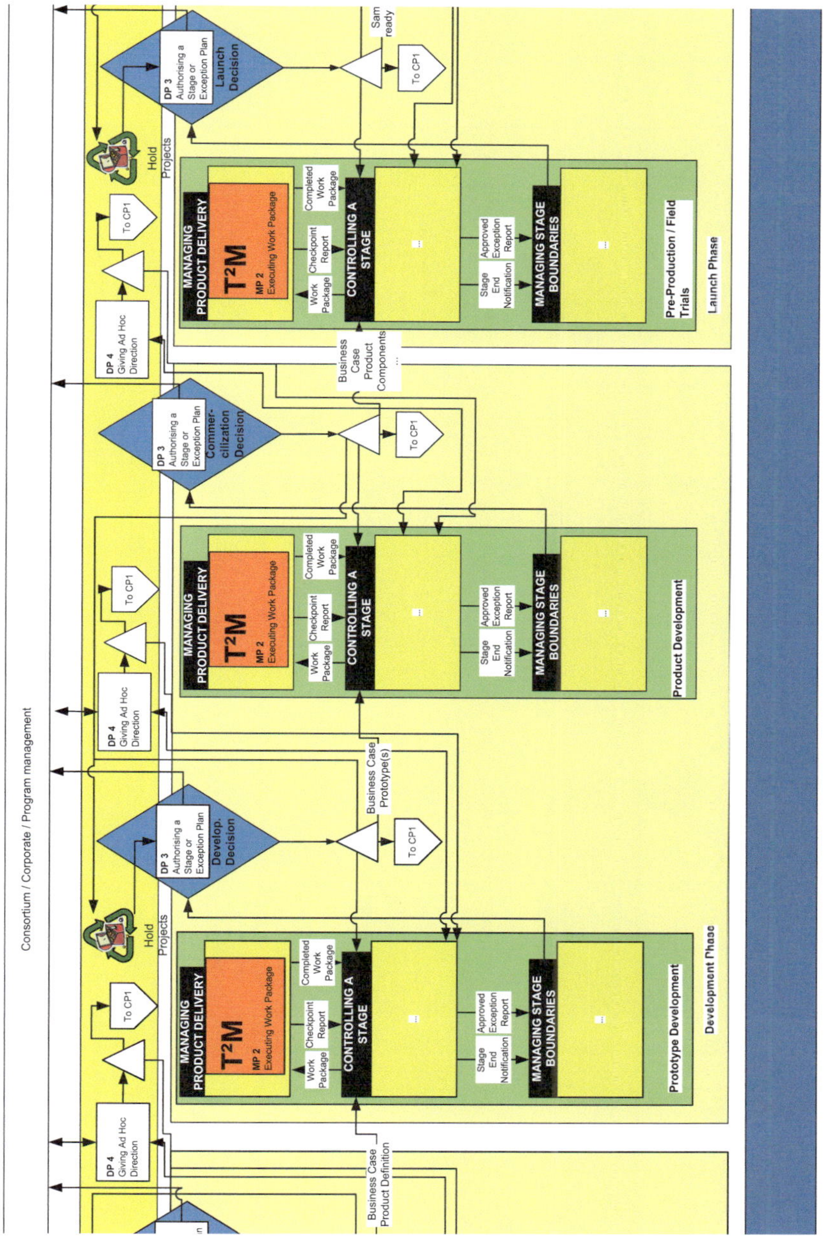

Fig. 5.6 Product engineering method with medium detail *middle* part for overview

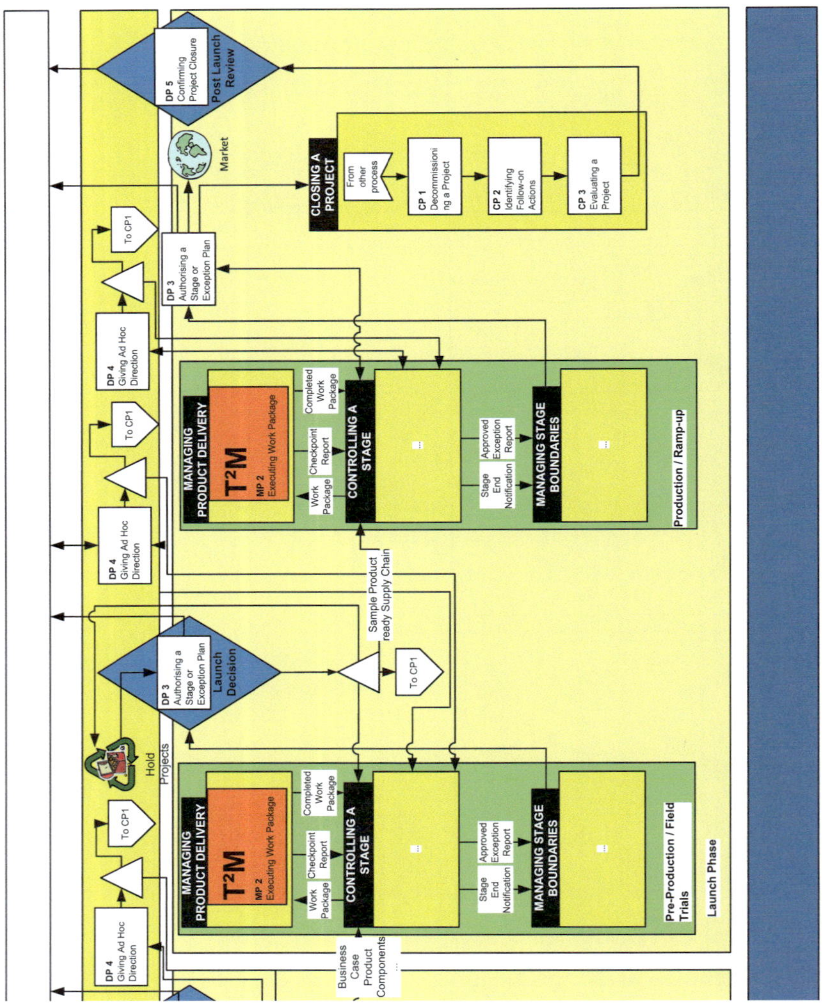

Fig. 5.7 Product engineering method with medium detail *right* part for overview

Table 5.2 Deliverables of idea capturing

Deliverable	Compulsory	Optional
Idea description	X	
Invention report		X

methodology as it will be repeated indefinitely, runs independently from all other engineering efforts, and its deliverables are informal at best. Nevertheless, it provides the starting point of any product engineering process and without a systematic approach for idea capturing many business opportunities might be missed. For formal reasons every captured idea becomes a deliverable. In the context of this methodology idea capturing consists of the following four components that are revisited on every iteration of the idea capturing stage:

The first component is the *Identifying Idea Sources*. The sources for new ideas within a company are typically the various research and development (R&D) groups as well as the marketing and sales (M&S) groups. These sources need to be identified and included in any idea fostering activity. Especially the input and feedback from the market via the M&S group is essential.

Establishing a Focal Point is the second component. The focal point is in charge of managing the idea generation process. This includes the stimulation of the idea sources (see above) as well as handling the idea generation process. In most cases the focal point will be a senior executive such as the chief technology officer (CTO), the chief innovation officer (CIO), etc. or their respective offices.

To *Grease the Path* for innovation can be considered the third component. The creativity itself can be increased via several means. As an example an idea kit like TRIZ (e.g., [4]) or similar could be given to the identified sources. Another possibility is to set up brainstorming sessions (also including customer and current project partners), to participate in trade shows and conferences, etc. A summary of other tools to increase creativity is given on page 111 in [2].

Establishing an *Idea capturing and handling system* is the last and most important component in a system for handling the ideas that are in fact the Deliverables of the Stage. The most important features of such an idea capturing and handling system are to

1. allow easy capture/submission of ideas, potentially as an open discussion; the inventor should provide the deliverables mentioned in Table 5.2
2. have a defined screening group applying a defined criteria catalogue;
3. provide timely feedback to submitters in writing;
4. establish an idea vault for "hold" ideas;
5. establish some award for successful ideas.

Most of these features can be implemented using common office and groupware solutions. Forum systems might be especially useful for discussing and collecting all relevant input concerning a specific product idea. The system is moderated by the focal point and it is his responsibility to fill the innovation pipeline with promising new product initiatives. If an especially promising idea comes up, the focal point may invoke an unscheduled *Idea Screen Gate*.

Table 5.3 Idea screen must-meet criteria

Criteria	met Yes	met No	N/A
Does the product fit to the overall business strategy?			
Is the market need preliminarily confirmed?			
Might the potential product deliver unique customer benefit?			
Is there a reasonable chance of technical feasibility?			

Table 5.4 Idea screen should-meet checklist

Criteria	Threshold	Rating (pt. 0–5)
Degree of alignment with overall business strategy	3	
Degree of alignment with design portfolio strategy	3	
Degree of alignment with process portfolio strategy	3	
Degree of alignment with packaging portfolio strategy	3	
Degree of potential customer benefit	2	
Likeliness of technical feasibility	3	
Overall threshold	20	

5.3.2 Idea Screen Gate

The *Idea Screen Gate* is the place where the newly captured product ideas are assessed. As there are limited predefined Deliverables in the *Idea Capturing Stage* that may trigger the idea screen, this gate has to be performed on a regular basis to assess several ideas / inventions in a single gate meeting. Alternatively it can be explicitly triggered by the focal point.

The *Idea Screen Gate* focuses on criteria like strategic fit, market attractiveness, technical feasibility, etc. The Gatekeepers of this gate are typically the CTO / CIO / VP Technology and one responsible representative of the individual design groups (at least electronics, MEMS and packaging) and one representative of the marketing and or sales department. Depending on the product idea, especially if it relies on a specific technology, additional decision makers should be consulted or included.

The first phase of the screening should filter out the not viable product ideas and is performed following the checklist presented in Table 5.3. Only products and project ideas fulfilling all the criteria with either "yes" or "N/A" will be assessed further with a scoring check. Product ideas failing the check are "killed" according to Stage-Gate® rules.

The scoring model check of the first gate is performed using the criteria checklist presented in Table 5.4. If the averaged value for one criterion is below threshold or the overall average threshold is not met, the product idea will be put on "hold" for later reassessment. An alternative course of action is to refer the idea back to the author in case the information provided in the invention report / idea description is not sufficient or not clearly expressed. In these cases the idea is moved to "hold" as well.

The gate does not assess new product ideas only (see Sect. 2.2.1 for a definition of new). But all ideas of status "hold" in the idea vault are reassessed periodically and might be moved to a different category depending on the results of the reassessment. This allows good ideas that previously were put on "hold" to be revisited and pursued at a later point in time. This might happen because the market, the strategy, or the technical feasibility has improved compared to the previous assessment. On the other hand ideas on hold might be killed completely because the market opportunities disappeared or because the product strategies have changed. The results of the scoring check are used to rank and prioritize the various competing ideas. Ideas passing the criteria check but not having a high enough priority will be sorted out into the idea vault in a "hold" state and reassessed later as well. The winning ideas will continue into the *Preliminary Investigation Stage*. The interaction of the gate and the business process is presented in Fig. 5.8. After the assessment of the new ideas the provider receives a written notification about the judgement.

5.3.3 Preliminary Investigation Stage

The *Preliminary Investigation Stage* is the first "real" stage adhering to the Stage-Gate® rules. Goal of the preliminary investigation is to briefly determine the product idea's technical, financial, legal/patent and marketplace merits and to evaluate in more detail whether the project strategically fits. Usually no more than ten to twenty person-days should be spent to perform a preliminary investigation of the product idea and the whole process should be finished within a month. Therefore one part-time resource from each of the relevant technical groups (depending on the idea classification, e.g., design, process, packaging group) will be assigned to the project. Additionally one part-timer from the M&S group, one part-timer from the legal/patent group and one part-timer from finance group will be assigned to the investigation.

The *Preliminary Investigation Stage* delivers several initial reports described below and listed in Table 5.5. They contain only preliminary data because this part of the project is executed in a short time with very limited resources. The reports are used as input in later stages / phases of the project where they are updated and extended.

The *Preliminary Technical Feasibility Report* contains first estimates regarding implementation technologies, development costs, development times, and potential risks. A proposal for a potential development plan should be contained as well. For the technical assessment a literature study and internal technical expertise and resources are used. The report collects information about existing processes or process steps creating a *Process Mapping*, integration schemes, initial designs and *Layouts*, *Packaging Alternatives*, and so forth usable for the new product idea. In detail the must-meet questions concerning the technical feasibility documented in Table 5.8 must be fulfilled.

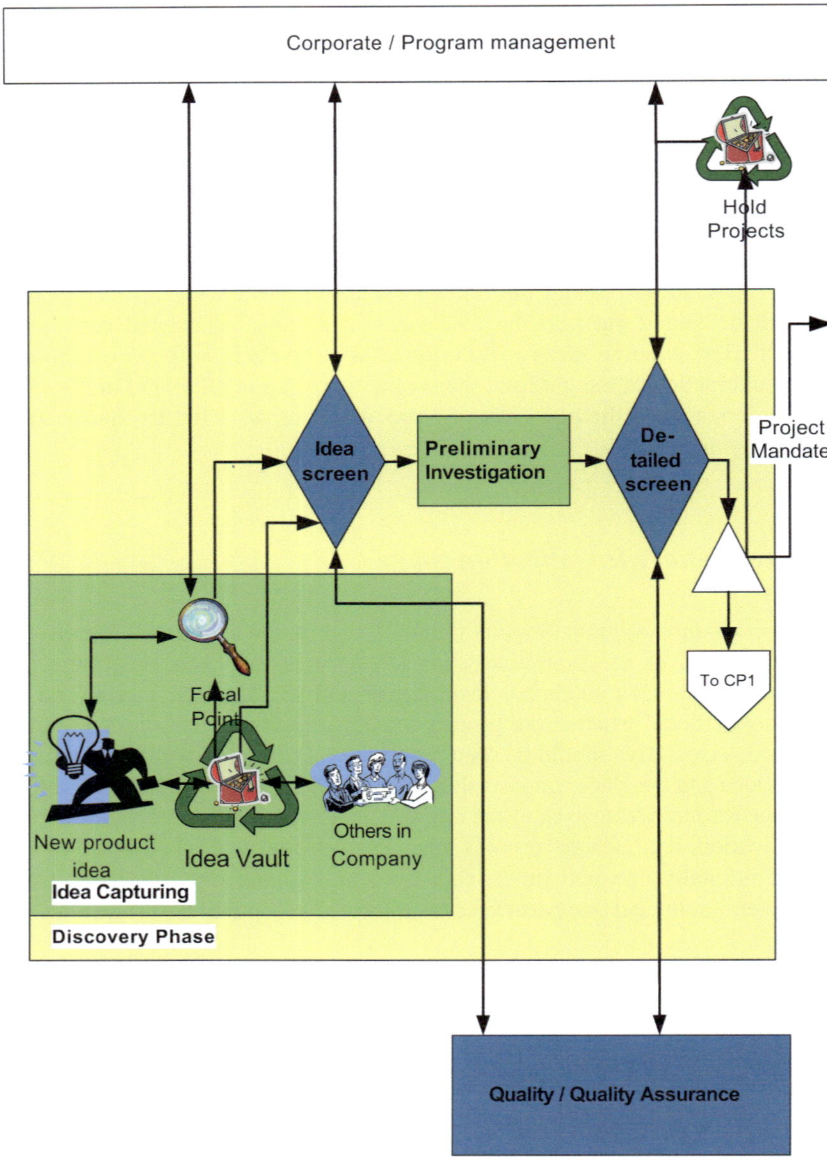

Fig. 5.8 Processes of the discovery phase

The *Preliminary Market Assessment Report* contains information about the first scoping of the market that includes assessment of market existence, estimated market size, market acceptance, and the product placement. The market opportunity window and competition should be briefly investigated as well. Because the

Table 5.5 Deliverables of the preliminary investigation stage

Deliverable	Compulsory	Optional
Preliminary technical feasibility report	X	
Preliminary market assessment report	X	
Preliminary prospects report	X	
Preliminary partner list	X	
Preliminary legal assessment report	X	
Draft project mandate (filled template)	X	
Project initialization stage plan (filled template)	X	
Req. for authorization to proceed	X	

Table 5.6 Preliminary partner list

Company	Role	Assets	Previous experience	Competitor yes/no	Contact yes/no

resources are very limited, the collection of this assessment is mostly based on preexisting market reports and internet resources.

The *Preliminary Prospects Report* contains an initial assessment on the potential product prospects. It takes the very rough estimates of potential sales (received from the market assessment), cost, required investment, and risk (based on technical feasibility assessment) into account. The report resembles the starting point for the creation of a Business Case in the later PE-Stages.

The *Preliminary Legal Assessment Report* contains results of an initial legal assessment. It covers relevant laws, regulations, and patents. Furthermore potential compliance requirements need to be evaluated and a framework to monitor those during the further development needs to be set up. Additionally the report needs to foster the awareness of the compliance implications on the decision maker and Gatekeeper level. Because in the MNT industry IP is a major factor, an initial patent check of the product idea is essential. The evaluation has to check whether there are blocking or competing patents in place and whether there are any negative legal implications developing a product based on the new idea.

The *Preliminary Partner List* contains summary information concerning potential development project partners. An outline of such a list is given in Table 5.6. This list is used in the following *Detailed Screen Gate* by the stakeholders to decide which potential partners to approach for a joined engineering undertaking.

The *Draft Project Mandate* summarizes the contents of the reports above. The main benefit of this report is that it suggests a list of potential project partners for the Product Engineering process. This list has been compiled from the information contained in the reports above and also contains a brief description of the benefits of including a specific partner in the Product Engineering process. A template for such a *Project Mandate* is described in more detail in Sect. 6.5.1. In case of a positive assessment during the *Project Startup Decision Gate*, the draft *Project Mandate*

Table 5.7 Responsibilities in the preliminary investigation stage (example)

	R&D management	Development management	Manufacturing management	Design team	Process team	Package team	Processing engineers	Operators	M&S	Legal	Financial	Project management
Preliminary technical feasibility report	A	I		R	R	R	S					I
Preliminary market assessment report	C								R			I
Preliminary prospects report	C								R			I
Preliminary partner list	C								R			I
Preliminary legal assessment report	C			C	C	C				R		I
Draft project mandate	C			C	C	C						R
Project initialization stage plan	C			C	C	C						R
Req. for authorization to proceed												R

will be extended / completed during gate execution. It is also used to initiate the *Startup Phase* of the PE method.

The *Initialization Stage Plan* lays out the schedule for the *Project Initialization Stage*. This schedule needs to contain all deliverables and tasks necessary to set up the product engineering environment including the *Electronic Project Binders* (please refer to Sect. 6.2.3 for more details). The generation of the plan follows the general planning guidelines described in Sect. 8.1.1.

On completion by the following gate the *Request for Authorization to Proceed* documents that the project is authorized to proceed into the next stage and grants all required resources.

A RASCI matrix for an example split of the responsibilities in the *Preliminary Investigation Stage* is presented in Table 5.7.

5.3.4 *Detailed Screen Gate*

The *Detailed Screen Gate* or *Project Startup Decision Gate* assessment is similar to the first gate but in the light of the additional information gathered during the *Preliminary Investigation Stage* and of course there is only one project idea under assessment. The assessment is also more thorough as there are two important and potentially costly decisions at this gate:

- Is the product idea worth to start up a development project?
- Which of the potential project partners should be contacted?

Table 5.8 Detailed screen gate must-meet criteria

Criteria	met Yes	met No	N/A
Does the product fit to the overall business strategy?			
Does the product fit to all business groups strategies?			
Is the market need preliminary confirmed?			
Is the addressed market of attractive size and is it growing?			
Might the potential product deliver unique customer benefit?			
Is there a reasonable chance of technical feasibility, especially?[a]			
Is a preliminary design study available?			
Is a draft manufacturing flow diagram available?			
Has an equipment capability study delivered promising results?			
Does the study of critical nodes of the process capabilities conclude only low or medium risks?			
Has the yield been estimated?			
Has a draft specification been created?			
Does a draft bill of materials exist?			
Has a study concerning inventory control been performed?			
Is a manufacturing execution system concept available?			
Is a capacity / throughput study available?			
Has the safety been studied?			
Are the shipping protocols scheduled?			
Is the training scheduled?			
Is the pricing / contracts under development?			
Have external vendors been established?			
Are there only minor or no legal/IP issues?			

[a] Extended by an extract of [7]

The assessment uses a must-meet list, presented in Table 5.8, a scoring model, presented in Table 5.9 and the *Potential Partners List*. Compared to the *Idea Screen Gate* both catalogues are supplemented with additional criteria.

The decision making during the Gate assessment is similar to the *Idea Screen Gate*. The must-meet criteria list is applied to the project idea, and if the project idea fails on one criterion, it is "killed." If the project takes this hurdle, it will be assessed with the should-meet checklist presented in Table 5.9. Project ideas scoring too low will be put on "hold" and transferred back to the idea vault. They have to pass another *Idea Screen* gate and go through another *Preliminary Investigation Stage* before being reassessed. If a project idea passes the assessment, the draft *Project Mandate* delivered to the gate will be finalized during the assessment session, filling, or at least deciding on, the remaining open items in the mandate. The finalized *Project Mandate* is used to initiate the *Project Initialization Stage* in the *Startup-Phase* via triggering the decided project executive and project manager with the *Project Mandate* as input. Additionally, the delivered *Stage Plan* for the *Project Initialization Stage* is amended and serves as an execution basis for the next Stage. All activities of the Gate are summarized in Table 5.10.

Table 5.9 Detailed screen gate should-meet checklist

Criteria	Threshold	Rating (pt. 0–5)
Degree of alignment with overall business strategy	3	
Degree of alignment with design portfolio strategy	3	
Degree of alignment with process portfolio strategy	3	
Degree of alignment with packaging portfolio strategy	3	
Degree of potential customer benefit	2	
Degree of potential customer need	2	
Degree of trade appeal and fit	2	
Degree of market size and growth appeal	2	
Likeliness of technical feasibility	3	
Appeal for patent portfolio	3	
Overall Threshold	28	

Table 5.10 Detailed screen gate activities with inputs and outputs

Inputs	Activities	Outputs
Prel. technical feasibility report	Authorizing a project initialization	Approved initialization stage plan
Prel. market assessment report		Approved project mandate
Prel. prospects report		Authorization to proceed
Prel. partners list		Potential partners list
Prel. legal assessment report		
Draft project mandate		
Initialization stage plan		
Req. for authorization to proceed		

5.4 Startup Phase

During the *Startup Phase* of the PE process a project consortium is built, the project is set up, and feasibility studies / proofs of concept are performed. At the end of the *Startup Phase* there should be a very clear idea about the product, its potential market impact, the role of each partner in the consortium, and the time frame of the development project.

As indicated in Fig. 5.3, the *Startup Phase* has two stages with very different areas of responsibility. These stages and their interaction are presented in more detail in Fig. 5.9. The *Project Initialization Stage* bundles the organizational tasks of setting up a project and the *Feasibility Study Stage* deals mostly with the technical and economical feasibility of the product (and of course requires a proper project frame to work). Therefore merging the two Stages makes only little—if any—sense. There is, however, the possibility to omit the *Feasibility Study* for projects where the feasibility of the product is for some reason out of question, e.g., when developing the next iteration of a product.

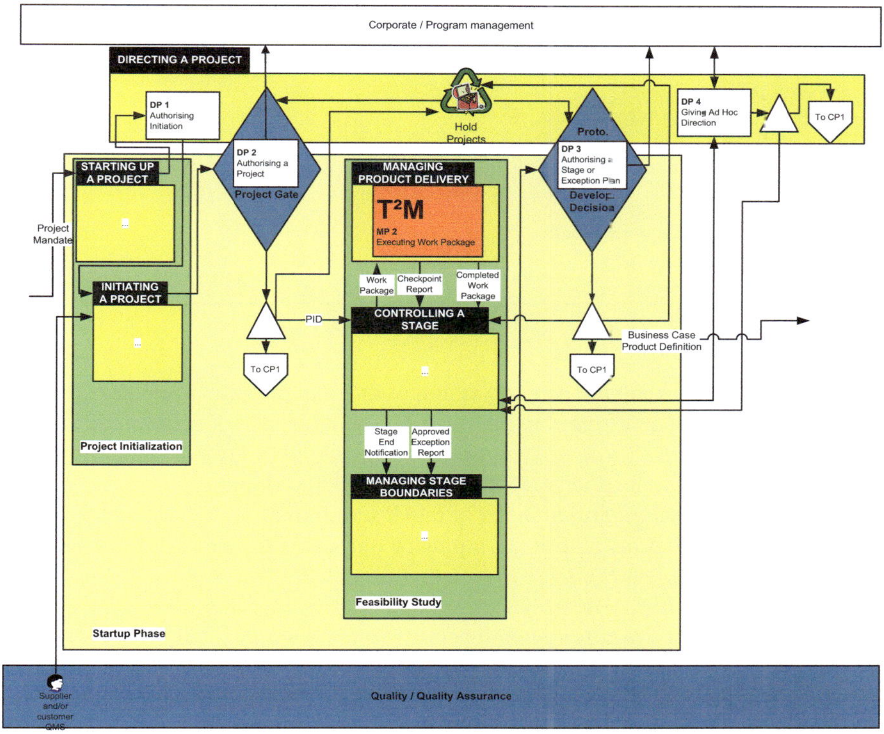

Fig. 5.9 Processes of the startup phase

5.4.1 *Project Initialization Stage*

During the *Project Initialization Stage* the project is initialized in accordance with PRINCE2® with some minor modifications that are outlined in the more detailed description in Sect. 8.1. The Stage is mostly a management and infrastructure Stage and can typically be performed within a short period of time (medium size projects approx. one to two weeks). Main purpose of the *Project Initialization Stage* is the initialization of the project and its infrastructure. This includes appointing responsible persons, setting up the project infrastructure, developing the *Project Brief* and *Project Initiation Document* out of the *Project Mandate*, setting up an initial *Business Case* and developing the various plans. The technical progress is limited to the definition of a vision for the final product as part of the *Product Brief*. For that purpose the *Project Mandate* and the *Preliminary Feasibility Reports* produced during the previous *Preliminary Investigation Stage* are used. The distinction between the *Project Brief* and the *Product Brief* is that the *Project Brief* is more concerned about the execution of the engineering project while the

Table 5.11 Project initialization deliverables

Deliverable	Compulsory	Optional
Project brief (filled template)	X	
Product brief (filled template)	X	
Draft project initiation document (filled template)	X	
Feasibility stage plan (filled template)	X	
Initial partner list	X	
Initial responsibility assignment matrix	X	
Stage end report (filled template)	X	
Req. for authorization to proceed	X	

Product Brief tries to capture an early vision of the final product. More details about the *Product Brief* can be found in Sect. 6.5.4. The sequence of actions follows the PRINCE2® standard for the process groups "Starting up a Project" and "Initiating a Project" and is described in more detail in Sect. 8.1. Additions and extensions of the PRINCE2® standard are included in the description as well. They are given by further explanations or as additional inputs or outputs, e.g., the addition of setting up an *Electronic Project Binder*.

The *Project Initialization Stage* delivers, apart from all the project management artefacts, intermediate results and project infrastructure documentation (e.g., *Project Brief*, *Risk and Issues Logs*, *Project Plan*, and *Project Quality Plan*, updated *Business Case*) and all the deliverables listed in Table 5.11. The high quality preparations of these deliverables are key assessment criteria in the next *Project Gate*.

The *Feasibility Stage Plan* lays out the schedule for the developments of the *Product Definition*, *Business Case*, *Product Requirements* and *Product Positioning Strategy*, *Legal Research Report* and *Risk & Issues Report*. It schedules the different work packages and activities to develop these Deliverables during the next stage. The generation of the plan follows the planning guidelines summarized in Sect. 8.1.1.

The *Initial Partner List* is very similar to the *Potential Partners List*. The main difference is that all partners on this list have agreed to participate in the project.

Another deliverable of this stage is the initial version of the *Responsibility Assignment Matrix* as defined in Sect. 2.3.4. At this stage it only contains information on the Stakeholders and Executive level. The Stakeholder level assigns responsibilities of project partners to general project activities like PRINCE2® management themes and Stage-Gate® stages and gates. The Executive level assigns executive personnel to key positions within the project (e.g., the Project Board, Gates, Transitions). The *Responsibility Assignment Matrix* is a living document that will be updated throughout the complete Product Engineering process.

A RASCI matrix for an example split of the responsibilities in the *Project Initialization Stage* is presented in Table 5.12.

Table 5.12 Project initialization stage responsibilities (example)

	R&D management	Development management	Manufacturing management	Design team	Process team	Package team	Processing engineers	Operators	M&S	Legal	Financial	Project management
Product brief	C								R			I
Initial partner list	C								R			I
Project brief	C								C	C	C	R
Draft project initiation document	C								C	C	C	R
Initial responsibility assignment matrix	C								C	C	C	R
Stage end report	C								C	C	C	R
Feasibility stage plan	C			S	S	S			C	C	C	R
Req. for authorization to proceed												R

5.4.2 *Project Gate*

The *Project Gate* assesses the project with the same criteria catalogues as the *Detailed Screen*, shown in Tables 5.8 and 5.9.

The Gate checks whether the overall situation has changed since the *Detailed Screen Gate* and additionally verifies the proper and quality execution of the *Project Initialization Stage* processes. Therefore it additionally checks and approves the creation of the *Initiation Stage Plan* (although already passed but checking for quality aspects), *Initial Partners List*, *Initial Responsibility Assignment Matrix*, *Product Brief*, *Draft Project Initialization Document (PID)*, and the *Feasibility Stage Plan*. Focus is also on the proper preparation of the overall *Project Plan* and *Quality Plans*.

The Gate is executed by the Project Manager and the Project Executive without higher level management involvement, so that it can be considered an intermediate Gate. Only if discrepancies between the two Gates concerning the evaluation of the Deliverables are discovered or a disagreement between the executive and project management is detected, senior management becomes involved in the gate execution. The outcome of the gate are similar to the *Detailed Screen Gate*. The project might be put on "hold" (very unlikely) or the *Feasibility Stage Plan*, *Initial Partner List*, *Initial Responsibility Assignment Matrix*, *Product Brief*, and the *PID* are approved. Recycles might occur in cases of incomplete or too low quality execution of the *Project Initialization Stage*. Table 5.13 shows the PRINCE2® process executed to perform the gate assessment together with the gate's inputs and outputs.

Table 5.13 Project gate activities with inputs and outputs

Inputs	Activities	Outputs
Initiation stage plan	DP2 authorizing a project	Approved feasibility stage plan
Project brief		Approved PID
Product brief		Approved product brief
Draft PID		Approved initial partner list
Initial partner list		Approved initial responsibility assignment matrix
Initial responsibility assignment matrix		Authorization to proceed
Feasibility stage plan		
Stage end report		
Req. for authorization to proceed		

5.4.3 *Feasibility Study Stage*

During the *Feasibility Study Stage* a more detailed *Business Case* for the product idea is constructed and a detailed feasibility analysis is performed. It can be considered to be the critical foundation stage where the technical, marketing, legal, and financial feasibility are worked out and the product is defined in detail. The stage consumes about ten to twenty percent of the overall project's development budget and therefore marks a considerable step forward on the way to a marketable product. The stage requires the participation of all involved business groups and subject matter experts. The outcomes must show that the product is feasible and that there is good evidence that the spending into the project will pay-off. This stage consumes a considerable amount of project resources and therefore requires a considerable amount of coordination. Because of that the previous stage has set up the project management framework and performed the initializing steps of the project management steps to prepare for a smooth start.

The *Feasibility Study Stage* is also critical for cooperative engineering as it is quite possible that for some partners the feasibility of the product development may not suffice to stay in the project. The bailout of project partners during this stage has therefore to be considered during execution and is a crucial part of risk management for the project. The stage performs many different activities, partitioned into several specialized work packages. Although those activities can be grouped into distinct technical and administrational areas, the developments are still performed in multidisciplinary/multi-partner teams because only this way the diverse perspectives of the different product aspects (technical, legal, marketing, etc.) are covered. The responsibilities and interactions of the project partners should be modeled and executed according to T^2M where the "real work" is planned and executed via the T^2M methodology.

Table 5.14 Feasibility study stage deliverables

Deliverable	Compulsory	Optional
Product definition (filled template)	X	
Business case (filled template)	X	
Requirements specification (filled template)	X	
Positioning strategy (filled template)	X	
Legal research report	X	
Risks&Issues report (filled template)	X	
Prototype dev. stage plan (filled template)	X	
Responsibility assignment matrix (filled template)	X	
Stage end report (filled template)	X	
Req. for authorization to proceed	X	

The Stage answers questions like "What is the product and who will it be sold to?", "Why to invest into this project?", and "How will it be undertaken; when, by whom, and how much will it cost?" Answers to those questions might require an update of the *Project Development Plan* that is handled by the "Finalizing a Stage" process steps. The technical, financial, legal, and M&S activities require coordination and management by the PRINCE2® processes and subprocesses.

The management activities of this stage are executed using the PRINCE2® standard processes. As presented in Fig. 5.9 processes from the groups Directing a Project, Managing Product Delivery, Controlling a Stage, and Managing Stage Boundaries are used. The detailed flow of these processes is the same for all R&D stages and is defined in more detail in Sect. 8.2. The resulting Deliverables of the stage are presented in Table 5.14.

In close collaboration with the M&S oriented-activities the *Feasibility Study Stage* performs detailed technical assessments. As a result of these assessments the *Product Definition* is delivered. Here initial lab work, e.g., material selection, process step selection, and package selection, product modeling and simulation, test designs, etc. will be performed to evaluate the technical feasibility in more detail. Even first prototypes might be produced to get a better feeling for the further efforts. The report collects information about the following activities aiming to supply Proof Of Concepts (POC) for the different areas / components of the product:

- existing processes or process steps become evaluated and selected, creating a matured *Process Mapping*
- the *Process Mapping* builds the baseline for the first steps of the *Process Development* activities with frequent changes in processes, materials, and equipment,
- different integration schemes for electronics, MEMS, and potential additional components are evaluated,
- conceptions of the designs and layouts are created,
- different packaging alternatives are specified.

For the technical feasibility check, the rule of thumb of spending about ten percent of the overall cost in this phase should be observed; otherwise, the majority of the engineering effort will already be done in this stage. It is the responsibility of the project manager to coordinate and control the right balance between meaningful feasibility confidence and over-engineering at this early stage.

The technical activities feed directly into the M&S activities trying to evaluate the market potential. Additionally, they are required for the assessment concerning the manufacturability at a customer site, and the necessary customer investments. The M&S activities feed back into the technical efforts to allow amendments to the concept and to better fit the product characteristics to the customer needs. Here the detailed modeling via T^2M plays out its advantages to allow for shared deliverables and close coupling of tasks of different responsibility groups. All these efforts are bundled into the *Product Definition* document. This definition can have varying contents depending on the application domain of the product. A more detailed description of the *Product Definition* is given in Sect. 6.5.5. For passing the following gate the Must-meet questions concerning the content of the *Product Definition* documented in Table 5.16 must be fulfilled.

Besides the technical activities, a more detailed financial analysis justifying the investment into the project is performed. The detailed financial analysis delivers a more concrete *Business Case* which will be enhanced in the next stages as the knowledge about the product, its market value, etc. increases. The financial analysis typically uses discounted cash flow methods, break-even analysis or return of investment calculation approaches.

Another deliverable of this Stage is the *Requirements & Positioning Strategy*. The M&S responsible persons at least have to perform a user needs-and-wants study, evaluating customer preferences, performance requirements and summarize them in the *Requirements Specification*. A template for such a specification is provided in Sect. 6.5.3 and is part of the *Electronic Project Binder*. Additionally a competitor screening needs to be done evaluating on their strengths, weaknesses, business performance, etc. A value-in-use analysis enables a comparison between the predicted own product characteristics and the alternative market solutions. The analysis allows also an estimate of the cost-benefit ratio from the customer perspective. Where applicable, first prototype presentations might be scheduled to get feedback from potential customers about the envisioned product characteristics. The results of these activities are summarized in the *Positioning Strategy Report*.

Based on the results of the technical feasibility analysis in combination with the *Preliminary Legal Investigation* from the *Discovery Phase*, a more detailed legal, regulatory, patent, and compliance assessment needs to be performed. For the deliverable *Legal Research Report* comparable approaches and patents need to be searched to prevent legal and IP issues. After fixing the approach and being sure about the feasibility of the product idea, first steps towards legally protecting the new product might be taken. This includes writing up new patent applications, registering new trademarks or copyrights, etc. All those activities and results need to be documented in the *Legal Research Report*, too.

Table 5.15 Feasibility study stage responsibilities (example)

	R&D management	Development management	Manufacturing management	Design team	Process team	Package team	Processing engineers	Operators	M&S	Legal	Financial	Project management
Product definition	R	R		C	C	C	S		C			I
Business case	C	C							C	S	R	I
Requirements specification	C	C							R	S	S	I
Positioning strategy	C	C							R	S	S	I
Legal research report	C	C							C	R		I
Risks & issues report	C	C							C	C	C	R
Stage end report	C	C							C	C	C	R
Responsibility assignment matrix	C	C							C	C	C	R
Prototype dev. stage plan	C	C		S	S	S			C	C	C	R
Req. for authorization to proceed												R

An explicit *Risk and Issue Report* is also necessary because the next Gate needs to take serious decisions and authorize massive investments into the project and the product idea. The report needs to list and quantify all open Issues and Risks contained in the *Issue and Risk Log* to address them explicitly. For all Issues and Risks a discussion about likeliness, impacts, workarounds, etc. has to be worked out.

The *Prototype Development Stage Plan* manifests the *Product Breakdown Structure* (PBS; defined is Sect. 8.1.1), the deliverable descriptions and the deliverable dependencies for the next Stage. It schedules the different work packages and activities to develop the product prototype(s). It plans the execution of the different market sampling, prototype test, refinement of the *Business Case*, and resolution of the legal issues. For that it follows the planning guidelines summarized in Sect. 8.1.1.

The deliverable *Responsibility Assignment Matrix* is an updated version of the *Initial Responsibility Assignment Matrix* produced in the previous stage. The initially defined information for the Stakeholders and Executives level becomes revised and additional roles and responsibilities become defined for the next stage. This deliverable is amended during every stage and therefore is an input and output document for all following stages and gates.

A RASCI matrix for an example split of the responsibilities in the *Feasibility Study Stage* is presented in Table 5.15.

Table 5.16 Prototype-development gate must-meet criteria

Criteria	met Yes	met No	N/A
Product still fits to the business strategy			
Product still fits into design portfolio			
Product still fits to the process portfolio			
Product still fits to the packaging portfolio			
Market need and profitable size confirmed			
Market requirements clearly defined			
Positioning strategy clearly defined			
Product Definition clear and agreed, especially[a]			
Is a design tested?			
Has a preliminary design FMEA been conducted?			
Is a preliminary manufacturing flow diagram available?			
Are statistical processing charts available as a draft?			
Have the equipment capabilities been tested?			
Are the tests of the critical nodes of the process capabilities positive?			
Has the yield model been created?			
Has a target specification been created?			
Does a preliminary bill of materials (BOM) exist?			
Has a demand forecast concerning inventory control been built?			
Is a manufacturing execution system demand forecast available?			
Is a capacity / throughput demand forecast available?			
Have the safety issues been identified?			
Are the shipping protocols identified?			
Is the training planned?			
Is the pricing / contracts under development?			
Have external vendors been established?			
Business case gives profitable outlook			
No major legal risks or issues			
Technical risks and issues are acceptable			

[a] Extended by an extract of [7]

5.4.4 Prototype-Development Decision Gate

The *Prototype-Development Decision Gate* marks the end of the *StartUp Phase* and leads directly to the *Development Phase*. Up to now only initial checks and feasibility tests have been performed. In the next stage a real prototype of the product will be developed. Therefore this Gate needs to be executed very thoroughly with higher management as well as customer involvement to reach sign-off on the *Product Definition*, the *Business Case*, and the *Development and the Marketing Plans*. Additionally agreement on the commitment of the required resources (capital, personnel, etc.) needs to be achieved. The priority shifts are reflected in the must-meet (see Table 5.16) and in the should-meet (see Table 5.17) criteria catalogues. The outcome of the Gate is similar to previous gates. The project might

Table 5.17 Prototype-development gate should-meet checklist

Criteria	Threshold	Rating (pt. 0–5)
Degree of alignment with overall business strategy	3	
Degree of alignment with design portfolio strategy	3	
Degree of alignment with process portfolio strategy	3	
Degree of alignment with packaging portfolio strategy	3	
Degree of market size and growth appeal	3	
Quality of requirements	2	
Appeal of positioning strategy	3	
Degree of potential customer benefit	2	
Degree of customer need	2	
Degree of trade appeal and fit	2	
Degree of newness of the product	2	
Clearness of Product Definition	3	
New Process knowledge captured in design and tracking system	4	
Profitability according to business case	3	
Level of risk (low risks→high # of points)	4	
Level of legal risks and issues (low risks→high # of points)	4	
Positive projects self-assessment	3	
High quality prototype development stage plan	3	
Overall Threshold	55	

Table 5.18 Prototype-development gate activities with inputs and outputs

Inputs	Activities	Outputs
Product definition	DP3 Authorizing a stage or exception plan	Approved prototype dev. stage plan
Business case		Approved product definition
Requirements specification		Approved business case
Positioning strategy		Approved responsibility assignment matrix
Legal research report		Authorization to proceed
Responsibility assignment matrix		
Risks&issues report		
Prototype development stage plan		
Stage end report		

be put on "hold" because of different prioritization, the *Feasibility Study Stage* might be repeated because of low quality or missing Deliverables or the project might be killed altogether because of various reasons (strategic fit, risks, issues, etc.). In case of a positive assessment, the *Prototype Development Stage Plan*, *Business Case*, and the other deliverables are approved and the project proceeds into the *Development Phase*. Table 5.18 shows the PRINCE2® processes executed to perform the gate assessment together with the inputs and outputs of the gate.

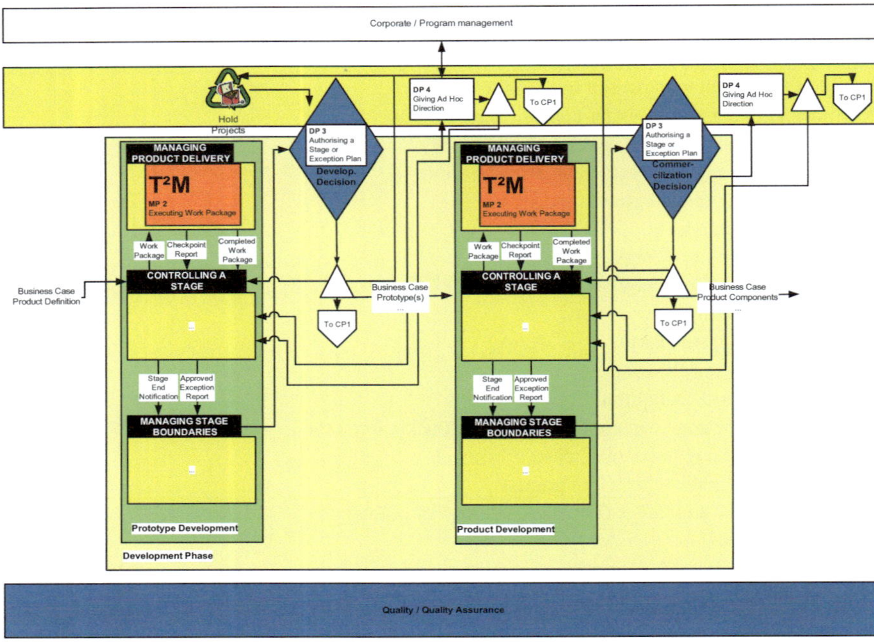

Fig. 5.10 Processes of the development phase

5.5 Development Phase

The *Development Phase* is, as the name implies, the main phase for product engineering. During this phase the product idea is refined into a concrete product ready for production, marketing, and sales.

The *Development Phase* is divided into the *Prototype Development Stage*, where a prototype of the product is evaluated and the *Product Development Stage* where the prototype is developed to its final state ready for mass production. Depending on the expected product engineering effort and the customer's wishes the *Prototype Development Stage* and the *Product Development Stage* can be merged into a single Stage.

The process of the *Development Phase* is depicted in Fig. 5.10. Both stages of this phase use the same principle management process as the *Feasibility Study Stage* that is described in more detail in Sect. 8.2.

5.5.1 Prototype Development Stage

The *Prototype Development Stage* is executed according to the *Prototype Development Stage Plan* as described at the *Prototype-Development Decision Gate* and follows the process described in more detail in Sect. 8.2. The main deliverable of

Table 5.19 Prototype development stage deliverables

Deliverable	Compulsory	Optional
Product prototype(s)	X	
M&S prototype report	X	
Updated business case (filled template)	X	
Legal resolution report	X	
Development stage plan (filled template)	X	
Responsibility assignment matrix (filled template)	X	
Stage end report (filled template)	X	
Req. for authorization to proceed	X	
Audit protocols		X

this stage is a prototype of the product that can be used to assess the product in context of its intended application scenario. For this reason the prototype has to be validated to a level that it passes in-house lab-testing and initial testing and basic quality assurance. The general idea being that the prototype can be handed over to the customer who performs the actual assessments in the application scenario and provides feedback from the user perspective. Even for internal developments it is advisable to engage preliminary customers for feedback, even if that means that additional effort is required.

For the prototype development efforts it should be noted that postponement and slippage is only allowed within the agreed project thresholds. Otherwise the project manager has to issue an *Exception Report* to the project steering committee which might assign additional resource or request additional effort. Serious project delays should be prevented under any circumstances because they might invalidate the *Business Case* or render the whole project useless. All threshold overshoots need to be treated according to the mechanisms documented in Sect. 8.2.4.

Because of the relatively long execution time of this stage, intermediate project or project management audits might be performed to ensure high quality standards of this stage, as the first assessment of the product by the customer may prove critical to the outcome of the whole project. Besides such extra efforts the stage uses the same principle management process as the *Feasibility Study Stage*.

During the *Prototype Development Stage* one or more *Product Prototypes* are developed. Such a prototype typically consists of several components representing one embodiment of the generated knowledge (e.g., physical device, software development kit, documentation). Selected variations of the prototype might be used to evaluate possible design, process, and packaging alternatives. In most cases the following activities are executed during this stage producing the deliverables listed in Table 5.19:

- the *Process Development* activities are continued to validate the manufacturing approach but with permissible changes in process, materials, and used manufacturing equipment,
- the selected integration scheme(s) for electronics, MEMS, and other components are verified and tested using initial components manufactured with the manufacturing processes still under development,

- designs and *layouts* for initial manufacturing are created,
- *Packaging Alternatives* are finalized and validated.

Depending on the project the prototype may also include a *Process Blue Book* together with processed prototype wafers and devices, designs for testing and qualification, fabricated test chips, and application notes with feasible and tested device integration and packaging schemes. Besides the defined prototype components a huge body of additional information is generated during this stage that has to be secured for the upcoming *Development Stage*. To collect this information in a structured manner a suitable infrastructure should be employed.

The deliverable *M&S Prototype Report* contains the results from constant market sampling and initial customer prototype test. The market sampling needs to carefully observe the marketplace for requirement changes or changes of the competitive situation. In cases of major changes a report based on the *Project Issue Template* has to be generated to make the project aware of the new economical constraints. These have to be taken carefully into account (see process CS3 in Sect. 8.2.4) and might require further escalation using the exception treatment steps(see Sect. 8.2.4). M&S may also issue a *Change Request*, using the CR template, to consider new requirements and their implications on product development. The *M&S Prototype Report* summarizes all M&S activities and needs to cover at least the identified issues, the change requests, and the results of the customer prototype tests.

The existing legal issues (law, regulatory, patent, etc.) need to be resolved also during this stage. Possible resolution to these issues typically includes filing additional patents to protect the new idea and buying licenses to prevent issues with infringed patents. The aim must be to finalize the resolution of all legal issues before proceeding to the next stage. The results of this resolution process need to be documented in the *Legal Resolution Report*. Especially unresolved issues or deadlock situations need to be addressed by the product engineering team, e.g., by developing workarounds.

As the knowledge about the project and the final product grows, initial assumptions contained in the *Business Case* can be substituted by more well-founded data about the market or the legal situation. Therefore the foundation data for the *Business Case* becomes more solid and the *Business Case* needs to reflect this fact. The updated *Business Case* is reassessed in the next gate to complete the projects and products picture.

In the following stage the knowledge generation for the various IP components and other parts of the product continues. In the deliverable *Development Stage Plan* the plans for this stage are laid-out. Of special importance is the prototype characterization according to semiconductor test procedures, the widening of the process windows for manufacturing, and the assessment results of the prototype chips. All deliverables of the next stage need to be considered in the *Product Breakdown Structure* (PBS), defined in Sect. 8.1.1, the *Deliverable Descriptions* and the *Deliverable Dependencies*. The plan needs to schedule the different work packages, activities, and tasks to develop the final product deliverables ranging from technical, legal, financial, and M&S deliverables. The generation follows the guidelines summarized in Sect. 8.1.1.

Table 5.20 Prototype development stage responsibilities (example)

	R&D management	Development management	Manufacturing management	Design team	Process team	Package team	Processing engineers	Operators	M&S	Legal	Financial	Project management
Product prototype(s)	A	C		R	R	R	S	S	C			I
M&S prototype report	C	C							R	S	S	I
Updated business case	C	C							C	S	R	I
Legal resolution report	C	C							C	R		I
Positioning strategy	C	C							R	S	S	I
Stage end report	C	C							C	C	C	R
Responsibility assignment matrix	C	C							C	C	C	R
Development stage plan	C	C		S	S	S			C	C	C	R
Req. for authorization to proceed												R

A possible RASCI matrix to split the responsibilities in the *Prototype Development Stage* is shown in Table 5.20.

5.5.2 Development Decision Gate

After the *Prototype Development Stage* the focus of the gate assessments shifts from whether to invest into the project to "What is the progress?", "Is the project on track?". Additionally, the completeness, the "positiveness," and the quality of the deliverables move now into the focus of the assessments. Of course, the technical, financial, and strategic attractiveness of the product are still being reviewed but with reduced priority. These shifts are represented by a change in the must-meet (see Table 5.21) and should-meet (see Table 5.22) criteria catalogues.

The outcome of the *Development Decision Gate* is similar to that of the previous gates. The project might be put on "hold" because of changed priorities or the *Prototype Development Stage* might be repeated because of too low quality or missing deliverables. The project might be killed altogether as well because of various reasons (strategic fit, risks, issues, etc.). However, the likeliness to hold or kill the product is reduced because of the already high investments put into the project. In case of a positive assessment, the *Development Stage Plan*, the updated *Business Case*, and the other inputs are approved and the next stage is authorized. Table 5.23 shows the PRINCE2® process executed to perform the gate assessment together with the gate's inputs and outputs.

Table 5.21 Development decision gate must-meet criteria

Criteria	met Yes	met No	N/A
Product still fits to the business strategy			
Market outlook (size, need, positioning) still suitable			
Business case gives profitable outlook			
Prototype(s) of good quality, especially[a]			
Is a design tested?			
Has design FMEA been finalized?			
Has a preliminary Process FMEA been conducted?			
Is a tested manufacturing flow diagram available?			
Is the identification and the establishing of statistical processing charts initiated?			
Have the equipment capabilities been identified?			
Are critical nodes of the process capabilities meeting the physical specs?			
Has initial yield been achieved?			
Has the specification and the acceptance criteria been established?			
Has the bill of materials been finalized?			
Has the initial plan concerning inventory control been created?			
Are the models for the manufacturing execution system in development?			
Has the capacity / throughput been established and constraints been defined?			
Has a safety plan been initiated?			
Are the shipping protocols established?			
Is the training initiated?			
Is the pricing / contracts under development?			
Is the qualification of external vendors ongoing?			
Legal risks or issues resolved			

[a] Extended by an extract of [7]

Table 5.22 Development decision gate should-meet checklist

Criteria	Threshold	Rating (pt. 0–5)
Degree of alignment with strategies	3	
Quality of prototype(s)	4	
Degree of market outlook appeal	4	
Positive M&S prototype report	4	
New development knowledge captured in process development execution system (PDES)	4	
Profitability according to business case	4	
Level of legal risks and issues (low risks→high # of points)	4	
Audit reports positiveness	3	
Positive project self-assessment	3	
High quality development stage plan	3	
Overall threshold	38	

Table 5.23 Development decision gate activities with inputs and outputs

Inputs	Activities	Outputs
Product prototype(s)	DP3 authorizing a stage or exception plan	Approved development stage plan
M&S prototype report		Approved prototype(s)
Updated business case		Approved business case
Legal resolution report		Approved responsibility assignment matrix
Responsibility assignment matrix		Authorization to proceed
Development stage plan		
Stage end report		
Req. for authorization to proceed		
Audit protocols		

5.5.3 Product Development Stage

The *Product Development Stage* is used to transform all components (internal and external) for all involved product aspects and materials to their final state. This ranges from M&S plans (roll-out, distribution, support, etc.) and materials (product brochures, other collateral, etc.) to contract templates and an updated *Business Case*. The final technical deliverables contain elements like process manuals, characterization and reliability data, mask designs, etc. For the progression of the Stage execution the same constraints apply as for the *Prototype Development* Stage. This includes the remarks made about project delays and exception treatment as well as additional efforts on intermediate project and project management audits.

During the *Product Development Stage* all required *Technical Product Components* will be developed and the Stage results in the list of deliverables summarized in Table 5.24. For a complete MNT / MEMS product many different components are required, e.g.,

- a manufacturing process and the corresponding IP (+protection)
- a device / MEMS design and architecture and the corresponding IP (+protection)
- an electronic design for read-out, calibration, etc. and the corresponding IP (+protection)
- a software development kit and device drivers to integrate the product into its operating environment
- a package selection, or a detailed specification for nonstandard packages and the corresponding IP (+protection)
- performance indicators and qualification vehicles
- test and qualification data

The *Prototype Development Stage* created one or several product prototype(s). During the *Product Development Stage* the most promising prototype design

Table 5.24 Product development stage deliverables

Deliverable	Compulsory	Optional
Manufacturing process and process IP modules	X	
MNT / MEMS design IP modules	X	
Electronics design IP modules	X	
Packaging IP modules	X	
Performance indicators	X	
Qualification vehicles	X	
Test and qualification data	X	
M&S product components	X	
Confirmed beta-customer report	X	
Legal resolution report		X
Legal product components	X	
Responsibility assignment matrix (filled template)	X	
Stage end report (filled template)	X	
Updated business case (filled template)	X	
Pre-production/field trials stage plan (filled template)	X	
Req. for authorization to pre-production/field trials	X	
Audit protocols		X

is selected and developed further into the final product. Certain product types (especially high volume or high prize products) justify to initially continue the development of two or three prototypes and select the most promising one later during the *Product Development Stage* for continued development and stop the development of the other(s). To widen the process windows, additional virtual and real experiments need to be performed. Essential process IP modules delivered by these activities are the *Detailed Process Description* with the full recipe and process.

Performance indicators of the manufactured prototypes can be derived from diverse assessments and tests. They are required to prove the maturity and the thoroughness of the developments to customers. They only can be indicators because similar devices manufactured in a different environment can have slightly different properties. The required indicators must supply information about the characterizations of the process window for process and device variations (geometry or device type), wafer yield data potentially collected from different environments and setups as well as data about environmental and reliability test results.

The qualification vehicles are to support the different verification steps during product porting and integration efforts (if required). They cover material qualification, device (MNT and electronics) qualification, and so forth. Additionally diverse procedures, templates, plans, checklists and guidelines for the execution of technology porting, integration, and production ramp-up have to be set up.

The design IP modules consist of device and cell designs for the qualification vehicles and designs using the device(s) manufactured with the process IP. Additionally the design IP consists of design rule manuals and design kits allowing the usage of the design IP modules within customer implementations and the customers design flow. The design kits will consist of, e.g., simulation models, cell libraries, design rules, and test modules.

Table 5.25 Product development stage responsibilities (example)

	R&D management	Development management	Manufacturing management	Design team	Process team	Package team	Processing engineers	Operators	M&S	Legal	Financial	Project management
Manufacturing process and process IP modules	C	A		C	R	C	S	S	C			I
MNT / MEMS design IP modules	C	A		R	C	C	S	S	C			I
Electronics design IP modules	C	A		R	C	C	S	S	C			I
Packaging IP modules	C	A		C	C	R	S	S	C			I
Performance indicators	C	A		C	R	C	S	S	C			I
Qualification vehicles	C	A		R	C	C	S	S	C			I
Test and qualification data	C	A		C	R	C	S	S	C			I
M&S product components	C	C							R	S	S	I
Confirmed beta-customer report	C	C							R	S	S	I
Updated business case	C	C							C	S	R	I
Legal resolution report	C	C							C	R		I
Legal product components	C	C							C	R		I
Stage end report	C	C							C	C	C	R
Responsibility assignment matrix		C	C						C	C	C	R
Pre-production/field trials stage plan		C	C	S	S	S			C	C	C	R
Req. for authorization to pre-production/field trials												R

For passing the following gate the must-meet questions concerning the content of the *Technical Product Components* documented in Table 5.26 must be fulfilled.

During the *Development Stage* the M&S activities are continued and even intensified and generate the *M&S Product Components* deliverables. They have to plan, and partly already start, the roll-out, distribution, and support activities based on the *Business Case* and cost models. Additionally various service and support scenarios have to be worked out, prepared, and initiated. This includes the preparation of several presentation and product training documents. Depending on the product setup and product, new or additional customer service and support facilities need to be planned and prepared. Apart from those product brochures, customer briefs and other collateral material need to be designed before the *Technical Product Components* become available. One major task of the *M&S Product Components* of this stage is the careful planning of the field trials. For that beta-customers need to be selected, approached, and confirmed. Required deliverables are the agreements / contracts with the beta-customers summarized in the gate deliverable *Confirmed Beta-customer Report.*

After deciding about the business, cost, support, and service models the contractual framework for selling the new product can be created and become part of the deliverable *Legal Product Components*. These contract(s) need to cover, e.g.,

Table 5.26 Commercialization decision gate must-meet criteria

Criteria	met Yes	met No	N/A
Product still fits to the business strategy			
Market outlook(size, need, positioning) still suitable			
Business case gives profitable outlook			
High quality manufacturing process and process IP Modules, especially[a]			
Has a process FMEA been finalized?			
Is a finalized manufacturing flow diagram available?			
Are the statistical processing charts populated, set up and controlled?			
Have the equipment capabilities been validated?			
Do the critical nodes of the process capabilities have a Cp > 1?			
Has a baseline yield been established?			
Are the specifications in place?			
High quality performance indicators			
High quality test and qualification data			
High quality MNT / MEMS design IP modules, especially			
Is the design in the optimization phase?			
Has the design FMEA been compared to the process?			
High quality electronics design IP modules, especially			
Is the design in the optimization phase?			
Has the design FMEA been compared to the process?			
High quality packaging IP modules, especially			
Is the design in the optimization phase?			
Has the design FMEA been compared to the process?			
High quality qualification vehicles			
Has the bill of materials been frozen?			
Is inventory control in place?			
Are the models for the manufacturing execution system in place?			
Have the constraints been addressed?			
Has a safety been rectified?			
Are the shipping protocols in place?			
Have the training documents been issued?			
Is the manufacturing pricing in place?			
Are the external vendors qualified?			
High quality M&S product components			
High quality Confirmed Beta-Customer Report			
High quality legal product components			

[a] Extended by an extract of [7]

Table 5.27 Commercialization decision gate should-meet checklist

Criteria	Threshold	Rating (pt. 0–5)
Degree of alignment with strategies	3	
Quality of product component(s)	4	
Degree of market outlook appeal	4	
Positive M&S product components	4	
Positive legal product components	4	
New development knowledge captured in process development execution system (PDES)	4	
Profitability according to business case	4	
Level of legal risks and issues (low risks→high # of points)	4	
Audit reports positiveness	3	
Positive projects self-assessment	3	
High quality pre-production/field trials stage plan	3	
Overall threshold	41	

the definition of the contract contents, IP disclosure plan, license scope, contracts for the services and payments, maintenance agreements, and payment schemes. This framework needs to be prepared in parallel with the development to speed up the sales and roll-out process after the technical product components become available.

A RASCI matrix for an example split of the responsibilities in the *Product Development Stage* is presented in Table 5.25.

5.5.4 Commercialization Decision Gate

The *Commercialization Decision Gate* determines the rate of success of the product engineering activities and decides whether to commence with the commercialization of the product. The gate validates whether the product is still commercially viable (reviewing the updated *Business Case*), whether the product components are of high quality, and whether the customer accepts the product characteristics. It uses the must-meet criteria catalogue presented in Table 5.26 and the should-meet criteria catalogue (see Table 5.27) .

The central component is the checking of the completeness and "positiveness" of all stage deliverables, especially the *Confirmed Beta-Customer Report*. The plans and deliverables of the M&S crew are reviewed carefully as well to verify the completeness and thoroughness of the planning for the *Pre-Production and Field Trials Stage*. In the case of positive assessment, the gate converts the results of the *Product Development Stage* into a beta-product and initiates the *Launch Phase*. In cases of incomplete or low quality deliverables, the *Product Development Stage*

Table 5.28 Commercialization decision gate activities with inputs and outputs

Inputs	Activities	Outputs
Manufacturing process and process IP modules	DP3 authorizing a stage or exception plan	Approved product modules
MNT / MEMS design IP modules		Approved M&S product components
Electronics design IP modules		Approved beta-customer report
Packaging IP modules		Approved legal product components
Performance indicators		Approved stage end report
Qualification vehicles		Authorization to proceed
Test and qualification data		
M&S product components		
Confirmed beta-customer report		
Legal resolution report		
Legal product components		
Responsibility assignment matrix		
Stage end report		
Req. for authorization to pre-production/field trials		

might be repeated to improve the shortcomings. A killing of the project in that stage is very unlikely but could happen in cases of lost commercial or strategic viability, e.g., in case the market opportunity has shifted dramatically. Table 5.28 shows the PRINCE2® process executed to perform the Gate assessment together with the Gate's inputs and outputs.

5.6 Launch Phase

The *Launch Phase* is the final phase of the Product Engineering process. During this phase the supply chain, the production line, and the distribution channels are ramped up and the product is finally introduced into the market. Just like the previous phase, this is divided into two stages separated by an optional intermediate gate. In the *Pre-Production Stage* the production line is prepared and small volumes of the product are manufactured to be distributed to selected customers. The *Pre-Production Stage* is the most important stage for marketing and sales as the market is confronted first time with the "real" product. The results of this stage are evaluated at the *Launch Decision Gate* that decides whether to enter mass production. During the *Production/Ramp Up Stage* the production lines are ramped up to full speed and the product fully enters the market. The *Post Launch Review Gate* concludes the whole Product Engineering process with an final assessment. Besides assessing the

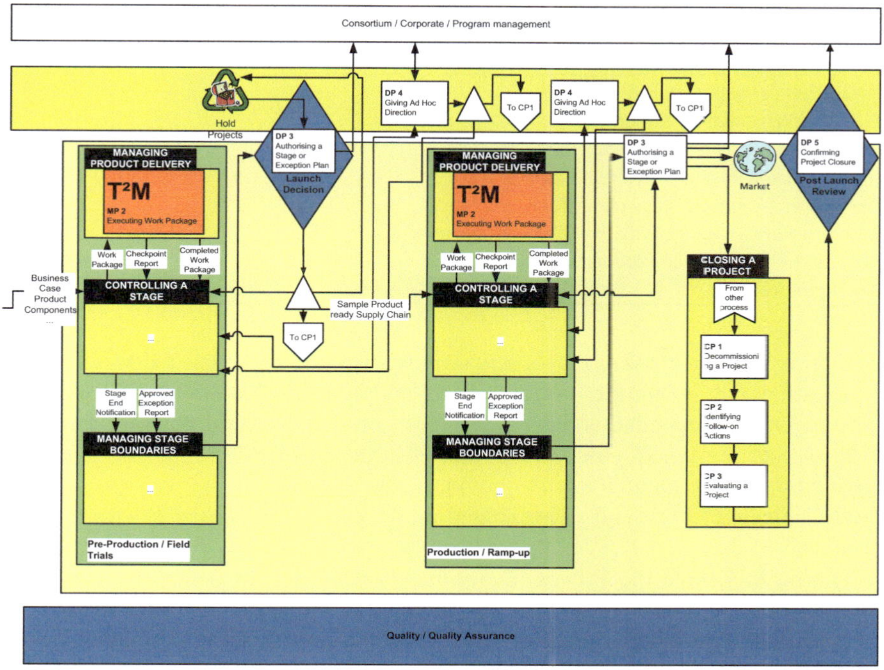

Fig. 5.11 Processes of the launch phase

completeness of the *Launch Phase* the *Post Launch Review Gate* is used to sum up the lessons learned during the engineering project for future reference. As such it is not really a gate in the sense of Stage-Gate®. There are no Go/Kill/Recycle decisions. After passing the *Launch Decision Gate* it is sure that the product will be brought to market. The *Post Launch Review Gate* is usually performed some time after the product launch itself, as indicated in Fig. 5.11. After the *Post Launch Review Gate* the project is officially closed and the project team decommissioned.

5.6.1 Pre-production / Field Trials Stage

The *Pre-Production Stage* is the first stage of the *Launch Phase*. The main purpose of the stage is to prepare a "working" production and supply chain for the final product (e.g., semiconductor fab → packaging → testing → distributor) and to start actively marketing the final product by providing samples to selected beta-customers. At this stage field trials are performed to assess the product's viability, prospective sales volumes, etc. under real market conditions. At the end of this stage there should be a realistic estimate of the expected sales, production cost, and volume to be produced. The deliverables of this stage are summarized in Table 5.29.

Table 5.29 Pre-production / field trials stage deliverables

Deliverable	Compulsory	Optional
Production & supply chain report	X	
M&S market sampling report	X	
Stage end report (filled template)	X	
Updated business case (filled template)	X	
Production / Ramp-up stage plan (filled template)	X	
Responsibility assignment matrix (filled template)	X	
Req. for authorization to Production / ramp-up	X	
Audit protocols		X

During the *Pre-Production / Field Trials Stage* the technical work is focused on creating and testing the production and supply chain for the product. For that the involved semiconductor fabs (for certain products several partners might be in charge of individual components like electronics, MEMS, ...), package fabs, testing houses, packaging service providers, distributors, etc. prepare the required production capabilities and sizing (based on the current *Business Case* for the new product).

During this stage the *Process Development* activities are replaced by *Process Improvement* activities. From now on changes in the manufacturing process are strictly controlled, as the product is now being qualified and the production has to reach the "established" maturity level. The results of these activities are summarized in the *Production & Supply Chain Report*. The small volumes manufactured during this stage are mainly used for field trials and product sampling.

The M&S activities are focused on providing the selected beta-customers with the agreed volume of the new product. Therefore the activities need to closely cooperate and interact with the beta-customers to gain as much feedback as possible concerning required product volumes and quality requirements. All information needs to be carefully assembled into the *M&S Market Sampling Report* building the precondition for a successful market launch of the new product. Furthermore the *Business Case* becomes updated reflecting the newly acquired information from the field trials.

A RASCI matrix for an example split of the responsibilities in the *Pre-Production/Field Trials Stage* is presented in Table 5.30.

5.6.2 Launch Decision Gate

At the *Launch Decision Gate* the customer decides whether or not to launch the final product. This decision should be based on the assessments done during the *Pre-Production / Field Trials Stage*. Naturally this is the most critical gate of the whole product engineering chain as a failed product launch is not only extremely costly but also likely to seriously damage the reputation of the customer in the target market. For all involved parties this is the last point to withdraw.

Table 5.30 Pre-production / field trials stage responsibilities (example)

	R&D management	Development management	Manufacturing management	Design team	Process team	Package team	Processing engineers	Operators	M&S	Legal	Financial	Project management
Production & supply chain report		C	R	C	C	C	S	S	C			I
M&S market sampling report		C	C						R	S	S	I
Stage end report		C	C						C	C	C	R
Updated business case		C	C						C	S	R	I
Responsibility assignment matrix			C						C	C	C	R
Production / ramp-up stage plan			C	S	S	S			C	C	C	R
Req. for authorization to production / Ramp-up												R

The gate assessment uses the must-meet criteria catalogue presented in Table 5.31 and is mostly driven by the customer. Table 5.32 shows the PRINCE2® process executed to perform the gate assessment together with its inputs and outputs.

5.6.3 *Production / Ramp Up Stage*

The *Production / Ramp Up Stage* is the final stage of the Product Engineering process and should, if all previous stages have been performed carefully, run smoothly. During this stage the production and supply chain prepared in the previous stage is ramped up to the intended volumes defined by the *Business Case*. After sufficient filling and throughput of the supply chain the product fully enters the market. Like the manufacturing process in the last stage, now the design, process and package component development is replaced by a continuous improvement approach—the *Process Improvement* activities. Every improvement requires a full qualification before being introduced into the production process. Although this stage is not concluded by a real gate in terms of Stage-Gate®, the stage itself still uses the PRINCE2® process "DP3 Authorizing a Stage or Exception Plan" in case an exception occurs and guidance is required by the stakeholders and / or the project directors board. The produced deliverables are mostly following the PRINCE2® standard and are listed in Table 5.33.

A RASCI matrix for an example split of the responsibilities in the *Production / Ramp Up Stage* is presented in Table 5.34.

Table 5.31 Launch decision gate must-meet criteria

Criteria	met Yes	met No	N/A
Product still fits to the business strategy			
Market outlook (size, need, positioning) still suitable			
Business case gives profitable outlook			
Production & supply chain ready and tested, especially[a]			
Has the design been revised?			
Is the process FMEA in adjustment / revision?			
Is a finalized manufacturing flow diagram available?			
Are the statistical processing charts monitored/OCAP/Adjustment?			
Have the equipment capabilities been finalized?			
Do the critical nodes of the process capabilities have a Cp > 1.3?			
Is yield improvement underway?			
Are the specifications in place?			
Has the bill of materials been frozen?			
Is the inventory control adjusted when required?			
The manufacturing execution system (MES) is modified as necessary?			
Capacity is added when necessary?			
Is the safety in sustain mode?			
Are the shipping protocols frozen?			
Have the training documents been revised?			
Has the manufacturing pricing been adjusted?			
Are the external vendors qualified?			
Field trials successfully conducted			

[a] Extended by an extract of [7]

Table 5.32 Launch decision gate activities with inputs and outputs

Inputs	Activities	Outputs
Production & supply chain report	DP3 authorizing a stage or exception plan	Approved production & supply chain report
M&S market sampling report		Approved M&S market sampling report
Stage end report		Approved stage end report
Updated business case		Authorization to proceed
Production / ramp-up stage plan		
Responsibility assignment matrix		
Req. for authorization to production / ramp-up		

Table 5.33 Production / ramp up stage deliverables

Deliverable	Compulsory	Optional
Production / ramp up report	X	
Stage and project end report (filled template)	X	
Lessons learned report (filled template)	X	
Post-project review plan (filled template)	X	
Follow-up action recommendations (filled template)	X	
Project closure recommendation	X	
Audit protocols		X

Table 5.34 Production / ramp up stage responsibilities (example)

	R&D management	Development management	Manufacturing management	Design team	Process team	Package team	Processing engineers	Operators	M&S	Legal	Financial	Project management
Production / ramp up report		C	R		C		S	S	C			I
Stage and project end report		C	C						C	C	C	R
Lessons learned report	C	C	C	S	S	S	S	S	R	S	S	I
Post-project review plan	C	C	C						C	C	C	R
Follow-up action recommendations	C	C	C						C	C	C	R
Project closure recommendation												R

5.6.4 Post Launch Review Gate

While not exactly a real gate according to Stage-Gate®, the *Post Launch Gate* serves as the final point of accountability for the project consortium. The gate is usually performed some time after the product launch. It assesses and approves the *Post-Project Review Plan*, the *Stage and Project End Report*, the *Follow-Up Recommendations* and the *Lessons Learned Report*. The approved *Lessons Learned Report* becomes publicized within the company as well as, may be in a stripped down version, project consortium to document the learning, extract the best practices, and prevent mistakes from being repeated.

This step officially closes the project, takes the lessons learned, decommissions the project team, etc. The real purpose of this second step of the gate is to foster a culture of continuous improvement: "The focus is on continuous improvement—a learning organization—rather than on blaming the team and creating a culture of fear and retribution" [3].

Table 5.35 Post launch review gate activities with inputs and outputs

Inputs	Activities	Outputs
Production / ramp up report	DP5 performing project closure	Approved post-project review plan
Stage and project end report (filled template)		Approved stage and project end report
Post-project review plan (filled template)		Approved and publicized lessons learned report
Lessons learned report (filled template)		Approved follow-up plans
Follow-up action recommendations (filled template)		
Project closure recommendation		
Audit protocols		

According to [3] the second step of the Gate is executed by the "Closing A Project" PRINCE2® process and the final confirmation with the PRINCE2® process DP5. Its subject matter and purposes consist of three main elements:

1. Having performance metrics in place: These metrics measure how well a specific new product performed. For example, were the product's profits on target?
2. Establishing team accountability for results: All members of the project team are fully responsible for performance results when measured against these metrics.
3. Building in learning and improvement: When the project team misses the target, or when deficiencies occur, focus on fixing the cause—stop this from happening again—rather than putting a band-aid on the symptom or, worse yet, punishing the team.

Practically, the *Post Launch Review Gate* evaluates the performance of the product in a relatively short time after the launch date (e.g., after 1 year). The achieved results (e.g., sales, market share) are compared with the expected results. Deviations from the expected results are carefully analyzed to prevent future misconceptions. The gate assessment is mostly driven by the customer. Table 5.35 shows the PRINCE2® process executed to perform the Gate assessment together with the Gate's inputs and outputs.

References

1. Bentley, C.: PRINCE2 a Practical Handbook, 3rd edn. ButterworthHeinemann, Oxford (2009)
2. Cooper, R.G.: Product Leadership – Creating and Lauching Superior New Products, 1st edn. Perseus Books, New York (1998)
3. Cooper, R.G.: Perspective: The stage-gate idea-to-launch process update, what's new, and nexgen systems. J. Product Innov. Manag. **25**(3), 213–232 (2008)
4. Gadd, K.: TRIZ for Engineers: Enabling Inventive Problem Solving, 1st edn. Wiley, Hoboken (2011)

5. Hahn, K., Schmidt, T., Mielke, M., Brück, R., Ortloff, D.: Micro and nano product engineering using data management for silicon-based fabrication process development. In: 9th IEEE Conference on Nanotechnology, pp. 337–340. IEEE-NANO 2009. IEE Explore, New York (2009)
6. Hahn, K., Schmidt, T., Mielke, M., Brück, R., Ortloff, D., Popp, J.: Distributed product engineering for microsystems devices. In: Proceedings of the 17th International Conference on Mixed Design of Integrated Circuits and Systems (MIXDES), pp. 453–457. IEE Explore, New York (2010)
7. MEMSIndustryGroup: Technology development process template. http://www.memsindustrygroup.org/i4a/doclibrary/getfile.cfm?doc_id=373 (2012). Accessed 10 March 2013
8. OGC: Managing Successful Projects with PRINCE2, 4th edn. The Stationery Office Ltd, Norwich (2009)
9. Ortloff, D.: Product engineering for silicon based MEMS IP. PhD Thesis, University of Siegen (2006)
10. Ortloff, D., Popp, J., Schmidt, T., Mielke, M.: A customer-driven approach to product engineering of micro and nano devices – requirement analysis. In: Proceedings of the 13th International Conference on the Commercialization of Micro and Nano Systems, Copenhagen (2009). COMS 2009

Chapter 6
MEMS PE Infrastructure Components

Abstract The Product Engineering Methodology introduced by this book and described in Chap. 5 requires significant tool support for its processes to be efficient. Additionally these tools need to provide certain features that are required for a sufficient MEMS Product Engineering Infrastructure. This chapter describes the most essential features extracted during studying such as business environment. Additionally it provides a view from the business perspective, identifying the areas of concerns which need to be covered by different tools of such an infrastructure. From the business perspective, the different building blocks (a.k.a. components) are described step by step and some tools supporting these areas of concern are introduced.

One key conceptual requirement that applies to all Product Engineering Infrastructure Components is the protection of intellectual property rights. This is due to the requirement to support multi-site / multi-partner collaboration. This directly conflicts with the need for fluent communication between all engineering partners / sites, because all exchanged information is potentially confidential. This leads to two aspects that need to be considered by all components of a Product Engineering Infrastructure: Every action must be properly authorized and all communication and data exchange has to be fully traceable. The way how this fits to the deep customer interaction was one of the focus points in the CORONA project.

In a distributed engineering scenario with multiple project partners, each partner has a set of tasks to accomplish that is part of the overall Product Engineering process. On several occasions during an engineering project, it is necessary to know what work has already been done by the other partners. Without providing a status tracking tool to which all involved parties have access, all status information is shared in other ways. Mostly this would be done by phone calls and emails, so there is no direct traceability and the information probably reaches not all people that would need it. This leads to misunderstandings or delays. Therefore, a software framework for supporting communication and status updates between the project

D. Ortloff et al., *MEMS Product Engineering*, DOI 10.1007/978-3-7091-0706-5_6,

partners should be Product Engineering Infrastructure. It needs to be emphasized that this infrastructure needs to be a distributed component, so each partner keeps the control over his/her respective data.

For the further descriptions it should be noted that the possible tools typically address several areas of concern so that a sharp distinction is often difficult to achieve. This is due to the overlap of the different areas. Additionally it is required to look at an aspect from several perspectives. On the other hand this overlapping provides a pre-integrated environment so that less components need to be integrated to form a coherent infrastructure.

6.1 Required Infrastructure Features

The analysis of organizations involved in MEMS development revealed [7] several required features of a product engineering methodology and infrastructure. These features can be summarized into:

- Enable knowledge management, especially for the core IP parts by management of the knowledge acquisition and management of usage and access.
- Enable reproducibility of engineering efforts by complete version control covering all necessary areas, input/output traceability, and complete data acquisition.
- Empower efficiency by easy access, early problem detection, regular status, and so forth with a slim but sufficient, extendable, and scalable approach, a structured and well-defined methodology and limited overhead.
- Give or increase control about status, progress, and issues by the usage of planning & tracking, constant status update, and issue awareness.
- Ensure to be target focused by sharing and communicating a vision, clear definitions, targets and responsibilities and active issue tackling.
- Enable easy and efficient communication by shared access to information and knowledge and synchronous and asynchronous communication means. This is especially necessary for efforts conducted in a multi-site / multi-partner engineering scenario.
- Aid better overall coordination by project management and clear responsibilities.

This poses the major question "How to best achieve all this?" which can be briefly summarized by:

- Use structured, proven, and well-defined approach
- Always execute along the same procedures
- Do *not* cut corners
- Constantly be aware of target vision, current planning, current status, known and expected issues, responsibilities, etc.

The defined approach, the execution along the same procedures, and the prevention of corner cuttings are covered by the processes and procedures and integrated quality controls of the Product Engineering Method described in

Chap. 5. These later aspects need to be taken care of by the Product Engineering Infrastructure. These areas are explored in the following sections looking from the business perspective and partly introducing possible tool components addressing the needs.

6.2 Configuration Management

Configuration Management (CM) is one of the cornerstones of a Product Engineering Infrastructure because it is one of its baseline components addressing

- reproducibility, by adding version control
- control, by storing the changes and the authors of the changes and
- communication/coordination, by building a communication platform.

Additionally Configuration Management is a central component of the engineering methodology as outlined in Chap. 5.

6.2.1 Definitions

Several slightly different definitions of the term Configuration Management can be found in literature. The definition in this book is based on the quite generic definitions of Configuration Management in the ITIL best practices [5]. Further aspects have been added from literature, e.g., from [2]. Analyzing these definitions and following the definitions in the IT Infrastructure Library (ITIL),[1] it can be noted that Change Management, described in Sect. 6.3, and Configuration Management are closely related processes.

Although Change Management and Configuration Management are closely related, here they are described as separate building blocks. This is because both components are closely related to requirements engineering as well. Change Management is described in Sect. 6.3 and Requirements Engineering is described in Sect. 6.4.

While the definitions in the above citations are relatively generic and abstract, the term Configuration Management is defined in more detail for the Software development area. There are several aspects to CM which are not explicitly covered by these generic definitions. According to [2] Configuration Management for software development needs to cover the aspects of Configuration Identification,

[1]The ITIL is coordinated and published by the Office of Government Commerce of the United Kingdom (OGC) [5], and defines business process and components for IT Service Management. It can be considered as a reference for best practices in the area of IT support.

Configuration Control, Status Accounting Audit and Review. Additionally it needs to contain Build Management, Process Management, and Team Work.

Although these definitions are targeted for software development and source code management, the mentioned issues are generic and should be applied to all kinds of engineering efforts. It is secondary that for software developments the source code builds the deliverable and product itself. This is not necessarily the case for other product engineering areas. For the here considered MEMS / MNT engineering all issues raised above are 100 % valid. They apply to the documentations, test results, and so forth of an MNT product as well. While the first block of the above aspects is considered to be the core of CM, the issues of the second aspect block will be attributed to other components. Altogether these points build a central focal point for the overall definition of the Product Engineering Infrastructure.

6.2.2 Electronic Product Engineering Flow

To support communication in all the different Stages of the engineering process and in between, the Electronic Product Engineering Flow (EPEF) that gives a status overview for all parties involved in a project has been introduced. This software provides a technical solution to these conflicting exigencies by sharing a consistent view on the project status to all participants and guiding the communication in between them. As such, the EPEF represents a software implementation of the Task Transition Methodology (T^2M) presented in Sect. 2.3.3. The main goal of the EPEF is to guide through the Product Engineering process and provide an entry point of information retrieval for multiple project partners. With the help of the EPEF system, the current project status is visible to all partners. As every partner is responsible for checking its respective tasks, problems can be foreseen early and the overall status is transparently visible all of the time. Of course, the more parts of the Product Engineering process are directly covered by a global software solution (i.e., the EPEF), the greater the benefit for the overall project. As the development of a distributed system is a major technical challenge, the feature set was kept as concise as possible. To allow for later extensions and adaption of the partner's business cases, extensibility has been a main concern during the specification. In a distributed engineering process, data has to be exchanged between different sites and partners. What data is exchanged has to be defined manually by the partners of the engineering process. This data can be "published" by the different partners and can be accessed by the other partners of the engineering process, according to the access permissions. For the CORONA project, the mission of the EPEF is to fulfill the requirements of the modules for regular status distribution and having a common information shring environment in a distributed multi-site, inter-company setting.

6.2.3 Distributed Project Binder

In Sect. 2.3.3 the Electronic Project Binder has been presented as a tool for the Data Organization module for a Product Engineering Infrastructure. The Distributed Project Binder is a direct extension of this concept to support the data exchange and data distribution in an engineering project carried out by a group of companies.

As an innovation project in the MNT area involves several companies—with their own expertise and intellectual property—controlled, secure, and retraceable data exchange is an important need. The Distributed Project Binder shall provide a tool supporting these requirements. It must enforce access rights to all data items inside the binder. This means, it must guarantee that data is only accessed by instances with appropriate rights, independent if these instances are a department in the company or a partner. Besides the enforcement of access rights, the Distributed Project Binder shall log all actions done to the contents. For every access to the data the binder will log at least date and time, the user and the change itself. With this information the complete history of a data item can be retraced in detail. As a basis for managing the content of a data item and its changes, a revision control system will be best suited.

The media for exchanging the data is an important challenge in the development of the Distributed Project Binder. Mainly the exchange will be done by unsecure media, like the Internet. The Distributed Project Binder has to provide authentication of the entity adding, reading or changing data and has to secure the communication channel by cryptographic means. In addition, the binder shall provide different distribution media, like email, Internet, or offline media.

6.3 Change Management

6.3.1 Features Provided/Addressed

Change Management is another fundamental component of a Product Engineering Infrastructure because it is addressing

- reproducibility, by organizing change and the application of changes
- control, by assessing changes and
- communication/coordination, by communicating approval of changes.

Additionally Change Management and Change Control is a central component and technique of the engineering methodology as outlined in Chap. 5 and depicted in Fig. 2.7.

6.3.2 Definitions

Web sites like wikipedia give a quite generic definition of Change Management. Analyzing this definition and following the definition in the IT Infrastructure

Library (ITIL; [5]), it can be noted that Change Management and Configuration Management, described in Sect. 6.2, are closely related processes. Additionally those two are closely related to requirements engineering described in Sect. 6.4.

Following the ITIL definition of, e.g., [5], Change Management and Configuration Management can be distinguished in the following way. Change Management focuses more on the overall coordination of changes while Configuration Management focuses more on the tracking and individual changed configurations. Therefore the Change Management takes care of the big picture and uses the Configuration Management to control the changes.

The aspects of Configuration Management are described in Sect. 6.2 while the minimal necessary components of the Product Engineering Framework for Change Management are described below.

6.3.3 *Components for Change Management*

As introduced in the Definitions, Change Management focuses more on the general coordination of changes. These can be changes in the infrastructure, as described in Change Management process of the ITIL [5], or changes applied to a project or product. Therefore the change management needs to foresee mechanism for the whole life cycle of a product, managing changes applied during the development or the maintenance of a product. It needs to contain tool and process components. A minimal set is described in the following paragraphs.

6.3.3.1 Component: Change Request

To modify an already existing product or a product engineering project it is essential to provide a predefined Change Request Template. Such a template needs to cover, e.g., the Object to change, the general reason(s) for the change(s), a detailed description of the change(s), prioritization of the change(s). All necessary aspects to cover are described in Sect. 6.5.6. The template and the change requests produced using this template need to be included in the projects repository.

6.3.3.2 Component: Change Request Procedure

The purpose of the Change Request Procedure is to give the Product Engineering Infrastructure a predefined, easy to execute means to achieve controlled change of already internally or externally released deliverables or to products currently under development. Because of that it builds one component of the Change Management and creates the gateway to propose, evaluate, and execute requested changes. The Change Management together with the Configuration Management takes care of keeping the CRs and the different configurations as well as the requirements for a project synchronized. The Change Request Procedure provides a framework to efficiently manage the Change Requests and apply them to the targeted deliverable. The CR Procedure needs to tie in with other components of the Product Engineering Infrastructure.

6.4 Requirements Engineering

6.4.1 Features Provided/Addressed

Requirements Engineering (RE) and resulting Requirement Specifications (see Sect. 6.5.3) are the initiator of a product vision, are used for calibrating a product idea against the market needs, or are analyzing a market need to generate a product vision. It is one of the baseline components addressing

- *target focus*, by defining the goal,
- *communication/coordination*, by communicating the aims and the constraints and
- *control*, by early definition of the goal and its constraints together with defining an orientation point during the developments.

This baseline laying specification is stored in the "specifications/in" folder of the standard electronic binder.

6.4.2 Definitions

Requirements define what features a product should provide to the real world and define the products constraints. As defined in [4, 8] they can be considered as the part concerned with anchoring development activities to a real-world problem. If requirements are not defined or not properly defined, this typically leads to late delivery, false assumptions, inappropriate products, and so on. Additionally problems in the requirements definition cause change requests after a product is already under development, in roll-out or already in use.

An overview introduction to Requirement Engineering and its techniques can be found on, e.g., wikipedia while [3, 4] go into much more detail. Additionally the later two give much more insight into the social aspects, the sequences, procedures, and different aspects of RE.

As a summary it can be said that Requirements Engineering is concerned with eliciting, analyzing, and documenting the product requirements taking the whole product into consideration. This includes all types of deliverables belonging to a product, e.g., documentation, plans, physical products, and so forth. To document all requirements and constraints (nonfunctional requirements) for a product, a requirements specification is put together by marketing and sales personal. The generation of this specification is coordinated by the requirements engineering role.

This document builds the baseline product documentation and sets the expectations on all sides, the customers and marketing, the engineers and management. It includes a product overview, a glossary, the statement of the functional requirements and the operational constraints. A more detailed description of the structure of this requirement documentation is given in the section Requirements Specification (see Sect. 6.5.3).

6.5 Project & Product Management Templates

This part of the chapter summarizes the project and product management templates used by the product engineering method. In principle this set contains all management templates advised by the PRINCE2® standard with additions of further predefined Deliverables. They are arranged in an *Electronic Project Binder*, presented as "Management Information" in the drawings. The binder and project repository created contains a full set of project management templates tailored to the needs of the product engineering method. The contents of the templates proposed in [1, 6] are adapted and extended to the new scope. Additional templates have been added to cater for not covered areas. The set contains templates for all aspects of project management documentations with additions for the product documentations.

To limit the amount of descriptions and replications of PRINCE2® standard templates, the following sections only describe changed or added templates. The contents of the other templates cover minimally what is described in [1, 6]. All templates received additional standard sections for internal document management and quality assurance aspects. Added comments and hints ease the filling of the templates.

6.5.1 Project Mandate

The *Project Mandate* template is reproduced from the PRINCE2® standard and covers the aspects of Authority, Background, Project Objectives, Scope, Constraints, Interfaces, and Quality expectations. Additionally it outlines the Business Case (with reasons), the Project Tolerances, the Reference to any associated documents or products. Furthermore it gives an indication of the Project Executive and Project Manager and Customer(s), user(s), and any other known interested parties. Apart from that, the standard sections for internal document management and quality assurance aspects are contained in the template.

6.5.2 End Stage Report

The *End Stage Report* has been modified with additional explanations and a recommendation for the further course of the Project. In this the Project Team has to propose whether to go, recycle, hold, or kill the project. This recommendation summarizes the projects internal knowledge and should be the result of a critical self-assessment as input for the next Gate.

6.5.3 Requirement Specification

The one component laying the baseline of a product and its different versions is a *Requirement Specification*. To make the requirements collection process thorough and complete, a predefined template has been developed. It guides and supports the requirements elicitation and defines the different angles and perspectives required to evaluate on. The template covers the aspects of the purpose/objectives, defines the special terms, and defines the required features split into required, desired, performance goals, and nonfunctional requirements. Additionally it needs to evaluate on the justification&benefits, target customers, deliverables, and competitors&alternatives. Therefore the template replicates the following structure:

- Preface
  Providing a version history with change comments, document approval status, and an abbreviations table.
- Introduction
  To provide the product definition.
  - Summary
    Identification of the product by its name and a brief overview about it.
  - Purpose/Objective
    Desired outcome from the Project and/or Product Engineering goals.
- Terms
  Definition of all terms with unusual, new or special meaning.
- Features
  This is to address the "what."
  - Required
    List of required features.
  - Desired
    List of optional features.
  - Performance Goals
    Provides specific performance goals for the product
  - Nonfunctional requirements
    Provides a list of remaining requirements which are not requesting specific functionality or performance. This could be certain delivery mechanism, look-and-feel, time frames, development approaches, e.g., rapid prototyping.
- Justification & Benefits
  This is to address the "Why."

- Target Customers
  This is to address the "who" (e.g., Product Applications(s), Target Market, customer Names/Identification).
- Deliverables (with target dates)
  List of specific deliverables. These represent major quantifiable milestones with tangible results and a target for "When" the milestone must be complete. This includes an MS Project plan on an abstract/milestone level which can be used as a basis for the more detailed planning.
- Competitors & Alternatives
  This provides information on alternative implementations that should be or have been considered and competitor's products that serve as a point of comparison for effective product engineering.
- Comments & Assumptions
  This section documents any assumptions made and any additional comments. To specify any risks, constraints, and issues. Also, specifying any external requirements/dependencies.

This baseline laying specification is stored in the "specifications/in" folder of the standard electronic binder.

6.5.4 Product Brief

The vision statement or *Product Brief* is created during setting up an innovation project as part of the development Project Startup Activities in the *Project Initialization Stage*. It defines the product scope in terms of the development aim. Its main purpose is to define the vision for marketing and market assessment purposes. For a market-driven product the vision statement clarifies which requirements of the requirement specification will be addressed by the product and which are out of the scope of the development project. It clearly defines what should be achieved by the development project and gives an overview about the operation domain.

6.5.5 Product Definition

The *Product Definition* is a kind of an answer of the engineering organization to the *Product Brief* and *Requirement Specification*. It is developed after the initial concept assessments and prototyping tests are performed and defines in detail the characteristics, limitations, and implications of the product. The limitations and implications need to clearly address areas of not addressed requirements and implications from the internal perspective (e.g., required development resource) and the customer perspective (e.g., equipment required by customers, customers current investment focus). The *Product Definition* or specification defines the development target on a detailed level from several perspectives. For MEMS design and process IP products a minimum set of perspectives needs to cover

- Design data sheet
  explaining most important device/design characteristics of the electrical and structural domain and their interactions.
- Process blue book
  Description of the principle process flow, possibilities, alternatives
- Integration blue book
  Description of how device/process can be integrated/rolled-out.

6.5.6 Change Request

To modify an already existing or defined project or product it is essential to provide a predefined *Change Request* template. This template covers the aspects of defining the terms, the object to change, give general reason(s) for the change(s), give a detailed description of the change(s), prioritize the change(s), evaluate on the Implication(s) of the change(s), and discuss the pro(s) and con(s) of the change(s). In detail it contains the following content:

- Introduction
  Short introduction into the change wish
- Terms
  Definition of terms used in the document
- Object to change
  Specification of exactly which object should be changed. Only one object can be changed per CR. If several items of a project need to be changed, separate CRs need to be produced. An exception can be made, if a single change has implications to several items. Then all items, the change has implications to, need to be listed in the References and Implications chapter.
- General reason(s) for change(s)
  Some text describing the reasons of the change
- Detailed changes
  Detailed descriptions of all items which require change including detailed reasoning and proposals how to change it (approaches, etc). These can be "New requirements," "Extensions," "Adaptations," etc.
- Prioritization
  If the change request contains more than one severe change, a prioritization should be given to ease the determination of change planning.
- Implications
  Here a detailed description of the implications of the change/changes have to be given. This includes a discussion and evaluation of the pros and cons implementing the change versus not implementing the change.

- Discussion
  Here a detailed discussion of the pros and cons of the requested changes has to be given.

For reproducibility and communication reasons the template is included in the electronic project binder in the specifications folder.

References

1. Bentley, C.: PRINCE2 a Practical Handbook, 3rd edn. ButterworthHeinemann, Oxford (2009)
2. Berczuk, S., Appleton, B.: Software Configuration Management Patterns, Effective Teamwork, Practical Integration. Addison-Wesley, Reading (2002)
3. Gerald Kotonya, I.S.: Requirements Engineering – Processes and Techniques, 1st edn. Wiley, Hoboken (1998)
4. Nuseibeh, B., Easterbrook, S.: Requirements engineering: A roadmap. In: Proceedings of International Conference on Software Engineering, November 2000, pp. 35–46. ACM, New York
5. Office of Government Commerce: IT infrastruture library. http://www.itil-officialsite.com/ (2013). Accessed 24 January 2013
6. OGC: Managing Successful Projects with PRINCE2, 4th edn. The Stationery Office Ltd, Norwich (2009)
7. Ortloff, D., Popp, J., Schmidt, T., Mielke, M.: A customer-driven approach to product engineering of micro and nano devices – requirement analysis. In: Proceedings of the 13th International Conference on the Commercialization of Micro and Nano Systems, Copenhagen (2009). COMS 2009
8. Zave, P.: Classification of research efforts in requirements engineering. ACM Comput. Surv. **29**(4), 315–321 (1997)

Part III
Practical Method Application

While Part II presents the MEMS Product Engineering methodology in a very general and abstract form, Part III is all about the practice. Based on the real-world experience of the authors in the MEMS industry, a typical application scenario of the methodology is presented. The benefits and solution approaches of the methodology are demonstrated for several sample situations that are typical during the stages of Product Engineering.

Chapter 7
Application of the Methodology: A Sample Scenario

Abstract In this chapter a practical application of the methodology at the Institute of Electron Technology (ITE) (http://www.ite.waw.pl, 2013 [10]) is presented. The ITE is a major Polish R&D center in the field of semiconductor electronics and physics with a focus on development of innovative micro- and nano-technologies and systems, and their applications in microelectronics, optoelectronics, photonics, and micro/nano-systems. For technology qualification it is equipped with a MEMS/CMOS pilot production line. The real case scenario presented in this chapter is a special test scenario based on a former customer requested project and has mainly been executed within the EU project CORONA. As such, it does not follow usual ITE development procedures, but an adapted set of a specific combination of the product development method Stage-Gate® and the project management method PRINCE2® described in this book. Besides that, it is exemplary for most customer-driven development projects at ITE. The reuse of a former customer project experiences and technologies allows the direct comparison of the new CORONA-based methodology and the usual applied procedures during the product engineering process at ITE. The scenario on the basis of gathered experiences presented in this chapter reveals as a promising, novel approach for product engineering, especially in the context of MNT. Because of several licensing issues and NDA constraints some specific details of the design flow have to stay unmentioned but the main impact of the methodology should become visible.

7.1 Discovery Phase

The first step in product engineering is the development of a product idea, vision, or concept. If there is no product idea there is nowhere to start and even if there is an idea it might still be a bad one. The purpose of the discovery phase lies in capturing a project idea/vision and evaluating its feasibility.

In this particular scenario the customer does not have sufficient experience nor capabilities to successfully handle the full MEMS device development process.

D. Ortloff et al., *MEMS Product Engineering*, DOI 10.1007/978-3-7091-0706-5_7,

Therefore ITE has identified the already running automotive project as a good candidate for implementation of the new methodology and its validation. ITE has contacted other project partners of Corona project for support in methodology definition and its implementation into the real case project. One of the business models of ITE is to support customers from the initial idea of a MEMS device over the device specification to the development of a final product prototype or—in rare cases—even small volume manufacturing. The market launch of a product or medium / high volume manufacturing as a MEMS foundry is not part of the services offered by ITE.

The product idea in this example case is part of an innovative automotive diagnostic system that is still under development. Therefore it is likely that there will be several changes in application constrains and resulting specification. If the product development is successful, the chances are good that there are many future applications of the product idea in the automotive industry. The predicted lifetime should be similar to other automotive electronic systems currently in use.

7.1.1 Idea Capturing Stage

The product idea was identified by the customer as a small volume project. But taking into account the potential future application area, it was obvious that it had the potential to become a high volume product. When the customer first approached ITE there was already a detailed concept of the general system. Support was required mainly in developing the micro-mechanical and micro-electronical modules. The initial specification covered general design constraints like that a capacitive sensor principle should be used and constraints of the application like frequency range, nominal capacitance, bias and detected acceleration range.

The task of ITE during idea capturing stage was therefore to select an appropriate sensor design and a concept for the readout electronics that match the customer constraints, the constraints of the automotive industry, and—last but not least—the constraints of the design capabilities and manufacturing technology available to ITE. To achieve this, a quick analysis of the customer input has been performed that came to the conclusion that the best approach would be a MEMS vibration sensor with the intelligent electronics on board. In such a scenario the main objective for ITE would be the development of a complete, functional implementation of the specific vibration sensor as proposed by the customer. The practical application of the idea capturing stage covered:

- preliminary decision taken for optimal methodology regarding customer requirements,
- formulation of conclusions and coherent vision of the modular prototype.

It emerged from this stage that the product would incorporate two separate micro system modules: a micro-mechanical vibration sensor based on a capacitive comb-drive concept and an IC with the readout and calibration electronics.

7.1.2 Idea Screen Gate

The purpose of the idea screen gate is to decide if the product idea articulated in the idea capturing stage fits the business strategy, is technically feasible, and has sufficient market potential. The input to this gate is usually nothing more than a textual description of the product idea.

In the case of this project the input from the idea capturing stage was more specific than usual, since the customer approached ITE with a rather detailed system specification that gave a clear outline what kind of development activities were expected. Articulating a product idea for the MEMS components to be developed at ITE was therefore straightforward.

Results of the idea screen gate were that the project is very well in the scope of ITE's business model and that ITE is principally capable to perform the envisioned development activities. Additionally it was noted that the project for itself is of strategic importance as it may lead to the development of close, bilateral, and fruitful cooperation standards for effective and successful multi-site device development.

7.1.3 Preliminary Investigation Stage

This stage is devoted to practical verification of the design and fabrication potential. It also covers mitigation actions to assure the successful project conclusion. Purpose of the preliminary investigation stage is to reinforce that first impression with more detailed technical and economical assessments.

Besides the technical assessments, the stage covers mitigation actions to assure the successful project conclusion. For instance, dedicated backup solutions have been discussed for eventual implementation in crisis situations and multi-criteria business scenarios for the best, typical, and worst cases have to be created.

Some of the activities performed by ITE during this stage were:

- to verify the whole idea of the MEMS vibration sensor structure and periphery architecture,
- to test various fabrication process and technology variants,
- to test various fabrication process setups,
- to test the applicability of various materials to the particular application (sensor development).

Based on the knowledge gathered during this stage ITE became effectively the intermediate project partner serving as a link between the customer and the fabrication site. Additionally, several important issues of the product engineering setup have been considered during this stage. For instance, design and simulation works on the mechanical part of the prototype were performed locally at ITE. Software tools selected to be used for the design were: CoventorWare [7], Coventor

MEMS+ [7], HEDORIS [11], Coventor SEMulator3D [7], COMSOL [6], Process Relations XperiDesk [17].

A parallel development scenario has been chosen for the micro-mechanical part of the device. The development of the manufacturing did not take place within ITE but at an external foundry. All the fabrication knowledge like technologies, design rules, etc. had to be provided by the foundry. The communication with the pure play foundry, access to the knowledge base, access to design guidelines, and design rules has been provided by means of remote web access, during direct meetings, and teleconferences. Nevertheless ITE had to perform a lot of specific technology experiments on its internal fabrication line.

Concerning the IC circuitry ITE and its customer decided during the *Preliminary Investigation Stage* that it makes sense to contract ASIC fabrication to another project partner. Based on this finding at least three potential solutions would be available:

- one ITE internal solution based on ITE-CMOS technology
- one external at a foundry which offers commercial CMOS Multi Project Wafer (MPW) services
- usage of the MPW service of Europractice [8]

Initially ITE preferred to use commercially available production at an external fab. But as a result of the preliminary investigations ITE and its customer decided to use Europractice MPW service for ASIC fabrication. It is an applied standard in ITE that other companies/partners are contacted at the initial project stages for discussion of the project general idea.

7.1.4 Detailed Screen Gate

The purpose of the detailed screen gate is to decide if it is worth to start a product development project based on the results of the *Preliminary Investigation Stage* and to provide a *Project Mandate* as the input for *Project Startup Stage*.

To reduce the risk of a design failure, ITE decided to implement a two-step development approach for the electronic part of the prototype:

- In the 1st step the customer would design and fabricate the electronic components using a PCB board and commercially available elements in full collaboration with ITE and the external foundry involved in MEMS fabrication. This solution would serve as a test environment for the MEMS component.
- During the 2nd step the ASIC electronic part would be developed, simulated, and verified. The ASIC device fabrication in selected CMOS technologies is also scheduled in one of the available technologies supported by a commercially available MPW service.

In Fig. 7.1 the project scenario decided during the *Preliminary Investigation Stage* is presented. One concern was that due to the fabrication schedule and the

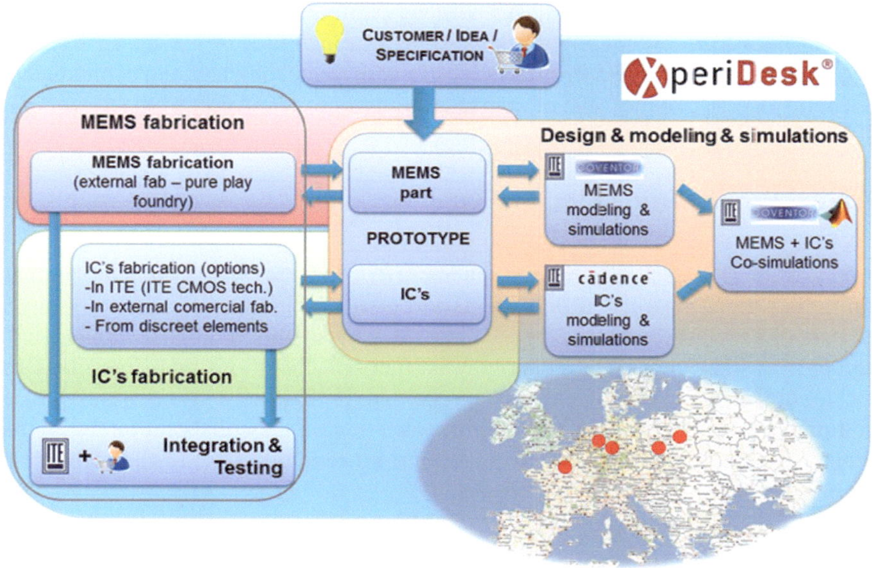

Fig. 7.1 Implemented real ITE scenario driven by customer [2]

standard design flow time-plan, it might be possible that fabrication of the CMOS module would not fit into the time frame of the project. Nevertheless even in the worst case scenario it would still be possible to validate the MEMS product using the PCB electronics.

It has also been noted that the customer intends to develop a modular system. The basic prototype of the system will only include a vibration sensor with intelligent electronics on-board for automotive applications, the general architecture of the system is that it will allow other types of sensors—pressure, temperature, acceleration, stress, etc. to be integrated as well. This aspect needed to be considered during the development activities.

7.2 Startup Phase

During the project *Startup Phase* a collaboration consortium is built, the project is set up and feasibility studies / proofs of concept are performed. At the end of the startup phase there should be a very clear idea about the product, its potential market impact, the role of each partner in the consortium, and the time frame of the project. In this particular case the product development efforts were part of an EU FP7 project; therefore, some of the issues mentioned have been agreed on in the initial consortium agreement and were fixed.

The *Startup Phase* has two stages with very different areas of responsibility. The *Project Initialization Stage* bundles the organizational task of setting up a project and the *Feasibility Study Stage* deals mostly with the technically and economically feasibility of the product (and of course requires a proper project frame to work).

7.2.1 Project Initialization Stage

Apart from the general funding scheme (EU, private, public–private partnership, etc.) there are several documents to be signed and agreements to be made as part of the *Project Initialization Stage*. In case of this project several specific agreements like NDA with technology partners or conditions and time plans for technological test runs had to be agreed on. Furthermore some basic project communication mechanisms were put in place to ease the collaboration during the project. One of these measures were setting up remote access to the technology knowledge base of the foundry partners.

7.2.2 Project Gate

The project gate assesses the project with the same criteria as the detailed screen gate. Additionally, it verifies the proper and quality execution of the *Project Initialization Stage* processes.

In this specific case all formal project activities have been—as mentioned—already agreed on beforehand as part of the planning of the EU FP7 project. The scope of the project is defined by work packages and individual partner responsibility is ensured by the work package leaders. The reporting scheme has also been fixed as part of the consortium agreement.

7.2.3 Feasibility Study Stage

During the *Feasibility Study Stage* a more detailed business case for the product idea is constructed and a detailed feasibility analysis is performed. In this particular case the feasibility studies were mostly concerned with evaluating the sensor concept for the MEMS component that was based on comb-drive structures with readout electronics. In strong cooperation with the customer ITE performed a study and an analysis of:

- Vibration sensor market, available products, and its specifications;
- Technologies and applied technological solution, especially construction based on the movable comb-drives;
- and with special attention to technologies available in the frame of CORONA project consortium.

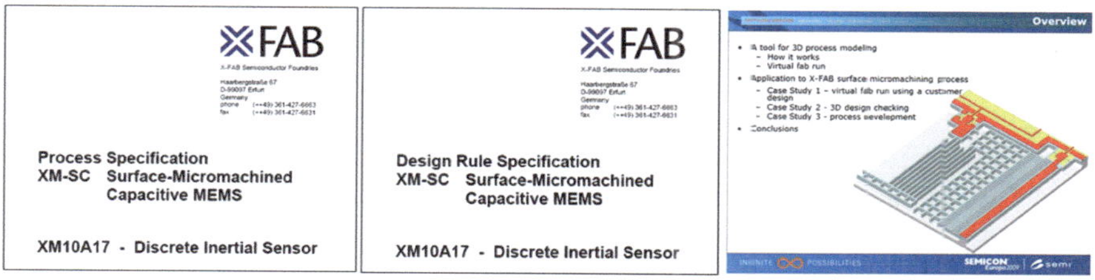

Fig. 7.2 X-FAB XM-SC technology description and sample design rule analysis [9, 18]

Based on those studies ITE selected the following available technologies for pre-studies:

- MEMS sensor
  - X-FAB XM-SC technology
  - ITE MEMS Technology
- MEMS readout
  - X-FAB XC10um CMOS X-FAB technology
  - ITE CMOS Technology
  - Other commercially available MPW service

Within the budget of the EU FP7 project X-FAB offered three fabrication runs of the XM-SC process technology. The detailed schedule of the cooperation had been discussed between X-FAB and ITE. Technological lines availability, project milestones, and design constraints also had to be taken into consideration.

The communication with X-FAB, access to X-FAB knowledge base, access to design guidelines and design rules is supported by means of remote web access to the X-FAB electronic knowledge base and during direct meetings and teleconferences. It is also important that within the project not only one partner was incorporated for the device fabrication, hence further cooperation for the standard development covers possible cooperation with another customer.

Before ITE decided to use X-FAB technologies made available for CORONA project partners by X-FAB the first step was to perform an analysis of the X-FAB shared knowledge base. ITE analyzed specifications and design rules of XM-SC group of processes made available by X-FAB. It was necessary to exchange a set of documents including specifications and presentations (Fig. 7.2). ITE participated in several discussions, teleconferences, and meetings. It was necessary to deeply understand all steps of the technology external for ITE. Main benefit of this intensive collaboration and the data exchange is the learning and understanding acquired by ITE for the case that ITE will have to perform external fabrication in the future again.

Nevertheless ITE had to perform many specific technological experiments on its own fabrication line. Main project goals of the internal ITE process evaluation were:

- to verify the whole idea of vibration sensor structure and periphery architecture,
- to test various fabrication processes and technology variants,
- to test various fabrication process setups,
- to test applicability of various materials to particular applications (sensor development).

The benefit was that ITE supported complex customer designs not only to be fabricated in the ITE fab but also to effectively exercising the cooperation with external fab with no risk of lowering the customer service quality or final device performance.

7.2.4 Prototype-Development Decision Gate

The *Prototype-Development Decision Gate* marks the end of the *Startup Phase* and leads directly to the *Development Phase*. Up to now only initial checks and feasibility tests have been performed. In the next Stage a real prototype of the product will be developed. As a result from the feasibility studies performed in the *Feasibility Study Stage* ITE decided to use the X-FAB XM-SC process as well as ITE's internal process for prototype manufacturing.

7.3 Development Phase

The *Development Phase* is the phase where most of the development work takes place. During this phase the product idea is transformed into a product ready for high volume production and product launch. The development phase is divided into the *Prototype Development Stage*, in which a fully functional prototype of the product is developed and evaluated and the *Product Development Stage* in which—based on the knowledge gained through prototype development—the product is brought to it final state.

For ITE the active participation in product development usually ends with the *Prototype Development Stage*. As a research institute, bringing products to high volume production is not part of the business model. After the *Prototype Development Stage* the responsibility for further development is handed over to either the customer or the manufacturing partner.

Depending on the complexity of the product and the customer's wishes the CORONA methodology described in this book allows merging the *Prototype Development Stage* and *Product Development Stage*. Such a general development stage is typical for development projects in which research institutes like ITE are involved, as the products developed are often just pilots for completely new product lines. The sample project described in this chapter is no exception from the rule. While the product idea has the potential to evolve into a series of very high volume products for the automotive industry, the main purpose of its first implementation is to gain an entry position into the market. For this purpose time to market, flexibility, and available features are more important than robustness, cost effectiveness, yield, and throughput. Therefore this project does not distinguish between prototype development and product development.

7.3.1 Prototype Development and Product Development Stage

The product has been designed to incorporate two separate micro-components: a micro-mechanical component that hosts the capacitive vibration sensor and an integrated circuit that contains the readout electronics for the sensor. The vibration sensor is based on the comb-drive sensor concept selected during the *Feasibility Study Stage*.

To fulfill the project work plan and increase the quality of the final product, ITE decided to apply two different development approaches in parallel for the micro-mechanical component. The micro-mechanical component has been designed using both the internal ITE manufacturing technologies and the external MEMS process offered by the play pure foundry (X-FAB). The customer has been involved directly in the development activities to ensure that the product not only meets the initial specification but also concrete needs of the customer. For lowering the manufacturing costs it was also planned to share the processing runs with another customer. By chance this customer was also a partner in the FP7 project. Had this not been the case additional precautions would have been necessary to make sure neither the project idea nor customer IP would be revealed to the other customer.

Due to high fabrication costs and time constraints the number of technological runs was very limited. During planning of the development stage it became clear that performing more than two runs in the limited time frame of the project would be very unlikely. Therefore extra effort has been put into stand-alone design and modeling activities to avoid as many potential design faults as possible before the first run. Several modeling & simulation cycles for various comb-drive-based structures have been performed (Fig. 7.3) at the beginning of the product development stage. The main computer assisted design (CAD) tools used were:

- CoventorWare, COMSOL—for design, modeling, and simulation
- AutoCAD, CoventorWare—for mask design.

Positive simulation results alone are usually not sufficient to ensure successful fabrication runs at ITE or at X-FAB—especially if completely new processing steps are involved. To limit the design expenses, selected technological experiments were planned and performed locally at the ITE facilities. The goal was to introduce some of the locally tested ideas/tricks also into the external fabrication process at X-FAB.

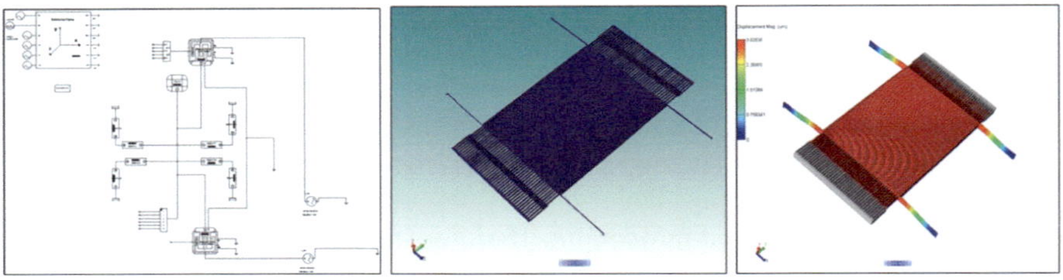

Fig. 7.3 Sample results of ITE design activities: behavioral model, 3D model and results of mechanical simulation of the comb-drive-based microstructure

7.3.1.1 Basic Sensor Development at ITE

One of the main challenges was that the manufacturing process of the micro-mechanical component has been developed from scratch: each processing step had to be carefully evaluated before the first processing run could be carried out at X-FAB. One of the most difficult processing steps developed at ITE was the release of movable structures. The approach chosen for that particular step is called "dry-release." This process step has to be carried out using a deep reactive ion etching (DRIE) tool. First, trenches are etched to the desired depth with simultaneous sidewall passivation which is a typical anisotropic trench etching approach (see Chap. 3). Once the desired trench has been formed, the passivation process is stopped. At this point the etch mechanism is changed from anisotropic to isotropic. By applying the isotropic etch the structural material (i.e., silicon) is underetched in the areas without passivation. By carefully adjusting the process parameters it has been possible to completely release the desired structures using this approach. Figures 7.4 and 7.5 show some of the test structures used during the development of the dry-release process at ITE.

Figure 7.4 shows the test structures used for the development of the dry-release etch: (A) is the full mask with all test-structures. (B) is the layout of a sample cell with various test structures in a more detailed view. An individual sample layout of a sensor structures is labeled (C). This particular sensor sample has a central perforated movable mass suspended on four springs (double-clamped beams) with two sets of interdigitated comb fingers located on opposite sides of the mass. In (C) the horizontal line indicates the location at which the device was cleaved for SEM (scanning electron microscope) inspection. A first trial version of the dry-release process was carried out on a standard 4" silicon wafer. The result of this trial is shown in Fig. 7.4 in insets (D) and (E) showing a cross-sectional view of the test structure. The pictures (D) and (E) have been obtained by SEM imaging.

The mask of another series of the test structures designed at ITE is presented in Fig. 7.5. The single sensor structure is shown on the right part of the figure. After the design, modeling, and simulation works the final structure designed for the first run in X-FAB was based on this design.

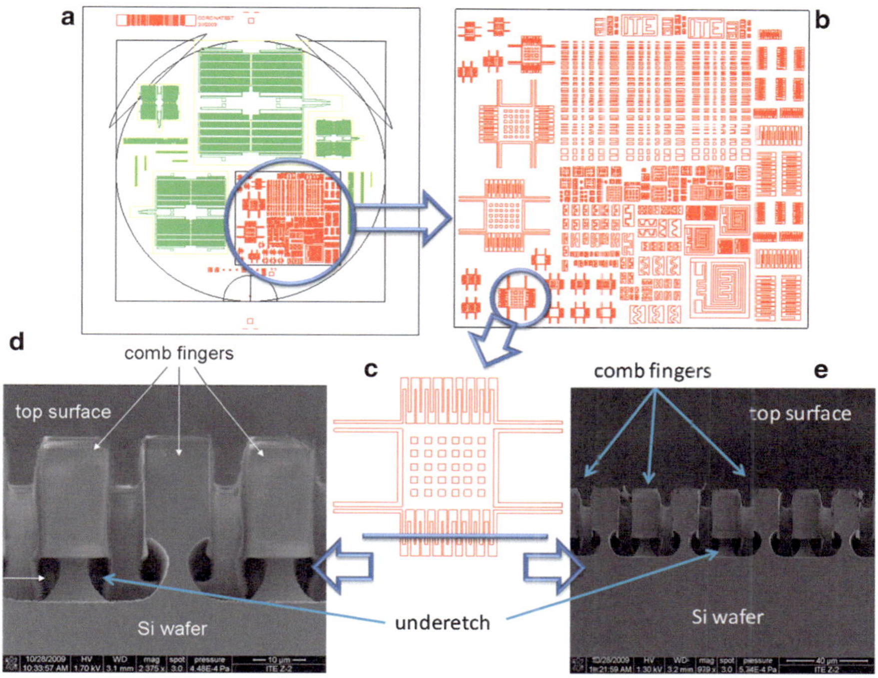

Fig. 7.4 Test structure layouts and results for release etch [3, 15]

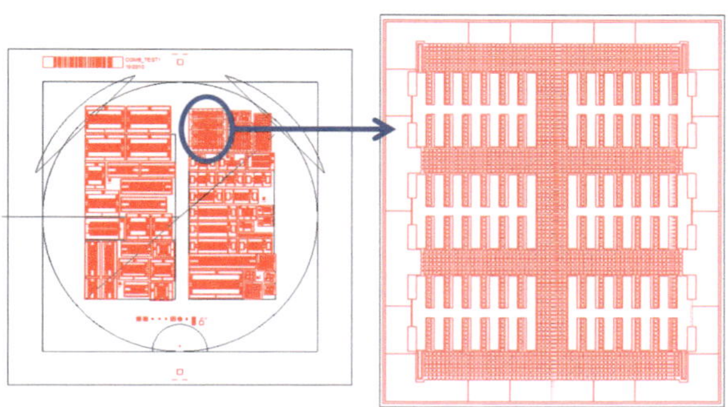

Fig. 7.5 Another set of the test structures mask (*left*) and almost final version of the sensor structure assign to X-FAB fabrication (*right*) [3]

After the preliminary experiments with test structures, the implementation and technology setup of the final structures of the vibration sensor had to be fitted to the target technology provided by X-FAB. Special attention was paid to the results of the preliminary technology experiments performed at ITE. The main part of the inertial sensor is a movable proof mass. It consists of two H-shaped parts connected at the middle section (Fig. 7.7). The movable mass is suspended on four U-shaped

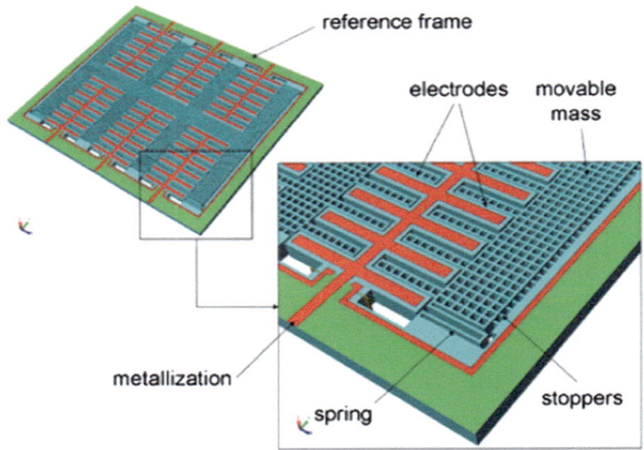

Fig. 7.6 Visualization of a central part of a MEMS sensor designed in ITE as a part of the prototype [3, 5, 13]

springs located close to its corners. Springs have been attached to the reference frame (Fig. 7.7). There are also comb-fingers perpendicular to the axis of the proof mass movement. This limits proof mass movement to only one degree of freedom, which makes it easier to analyze than sensors with two or three degrees of freedom. According to X-FAB design rules T-shaped metal electrodes have to cover every comb-finger. There are six electrodes in the sensor structure (three on each side) connected to the readout electronics.

The principle of sensor operation is based on proof-mass displacement. The sensor response to the external acceleration imposed to the structure is that the proof mass moves. The movement range is measured in relation to the reference frame. Inertial force causes the mass movement in a direction opposite to the direction of the acceleration imposed on the structure. The displacement is directly proportional to the magnitude of acceleration. Comb-fingers and proof-mass structure form variable capacitors with the capacitance being dependent on the proof-mass displacement. To find the exact parameters of the sensor behavior, the proof mass displacement was measured indirectly by detecting changes of above-mentioned capacitors formed by comb fingers.

The layout presented in Fig. 7.6 was created in the CoventorWare layout editor using a TCL scripting tool. This simplifies and speeds up the device adjustment process to reach the required parameters and serve specific features (like geometry) or to fit given design rules belonging to the selected fabrication technology. Additional analytical and FEM calculations have also been performed. After the MEMS part of the sensor design was finished, the generated layouts were sent to X-FAB for verification if the ITE design fulfills X-FAB foundry design requirements.

What's been not mentioned so far is the design of the read-out electronics. Similar to the MEMS sensor there have been several variations of readout

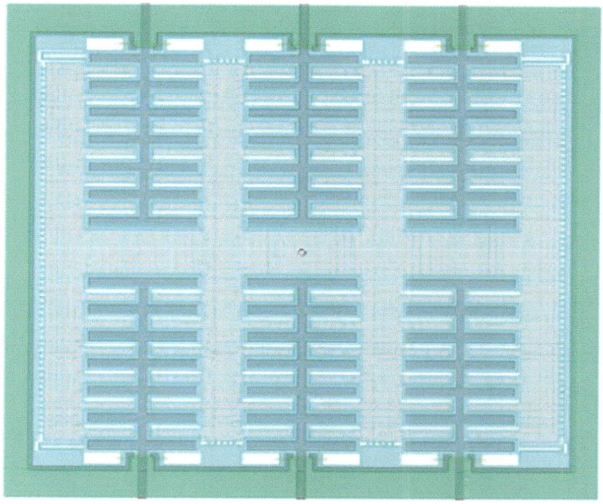

Fig. 7.7 Layout overview of a central part of a MEMS sensor designed at ITE [5, 13]

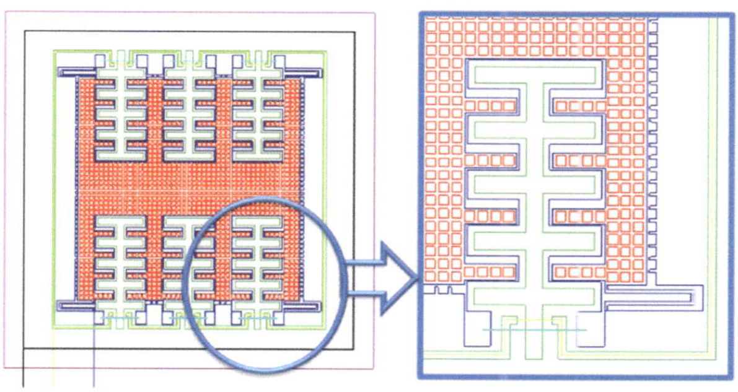

Fig. 7.8 Mask set transferred to X-FAB for verification

electronics developed at ITE. The readout electronics are fairly typical mixed signal designs that use a standard ic-manufacturing technology. A detailed description of those solutions would exceed the scope of this chapter and was briefly discussed in paragraph 7.3.1.5.

7.3.1.2 Preparation for 1st MEMS Technology Run at X-FAB

The test structure masks were written in GDS2 format (Fig. 7.8) and then sent for validation to X-FAB just after the preliminary design was finalized. The whole design successfully passed the verification step performed at X-FAB (Fig. 7.9) with only two minor changes required.

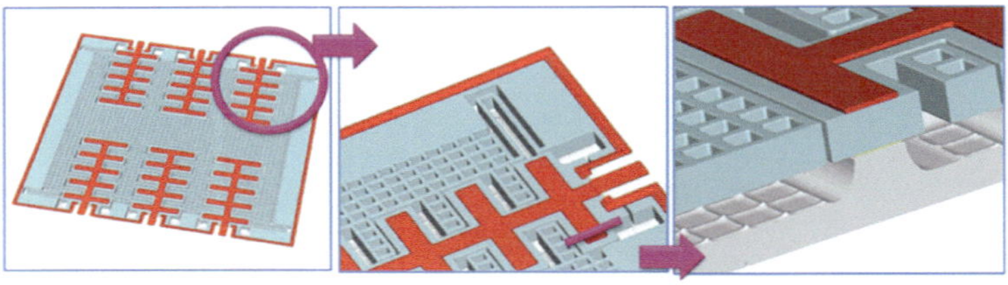

Fig. 7.9 Result of verification and visualization at X-FAB [2, 3, 5]

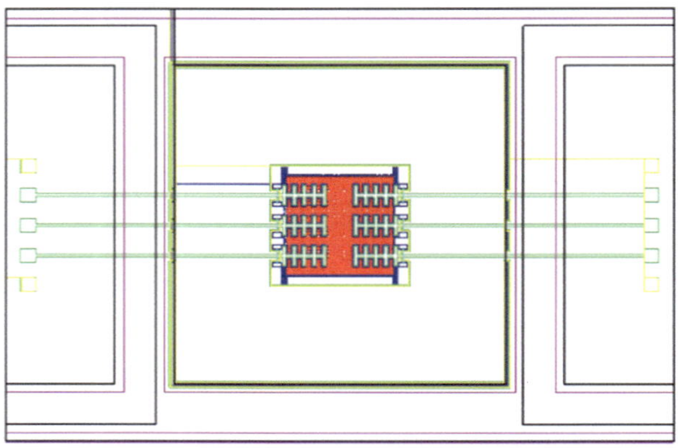

Fig. 7.10 Final version of the ITE design for fabrication in X-FAB [1–3, 5]

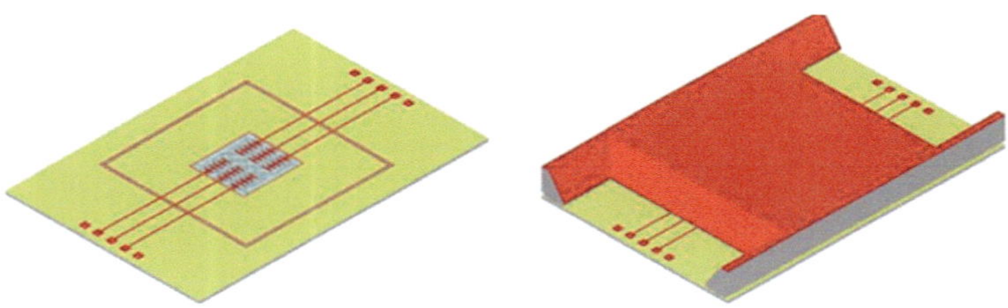

Fig. 7.11 Result of verification and visualization of the designed inertial sensor: device wafer (*left*); complete chip (*right*) [1–3, 5]

After applying the changes a corrected design was sent to X-FAB for repeated verification. Some missing elements have been added, others have been modified (Fig. 7.10). The result of verification and visualization by virtual manufacturing (Coventor Semulator3D) is shown in Fig. 7.11 and is shown step-by-step shown in Table 7.1.

Table 7.1 X-FAB virtual manufacturing run with Coventor SEMulator3D of ITEs design for 1st run (full chip) [1, 2, 7, 9]

Step 1—SOI wafer

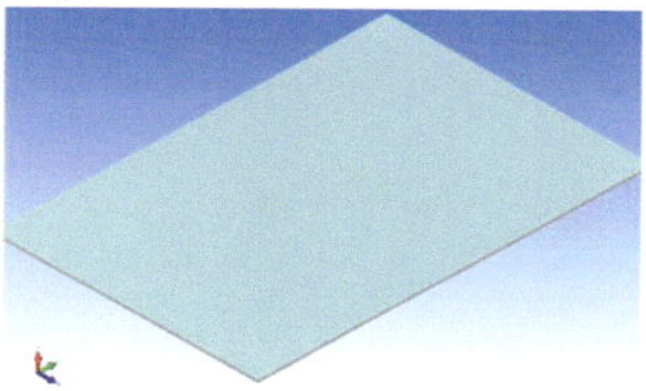

Step 2—Etch and refill isolation trench

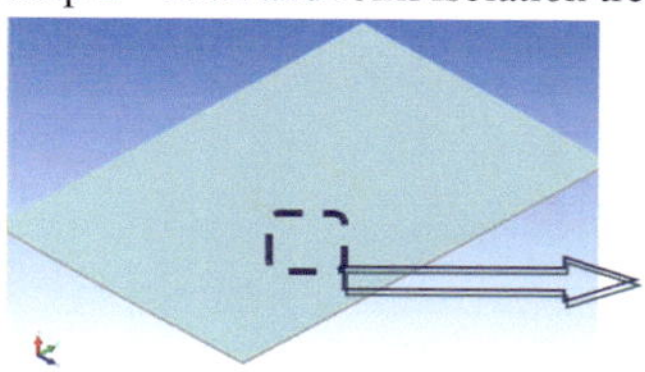

Step 3—Intermediate isolator

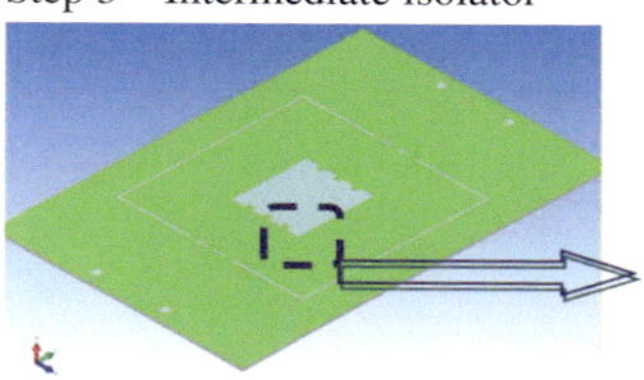

Step 4—Metallization

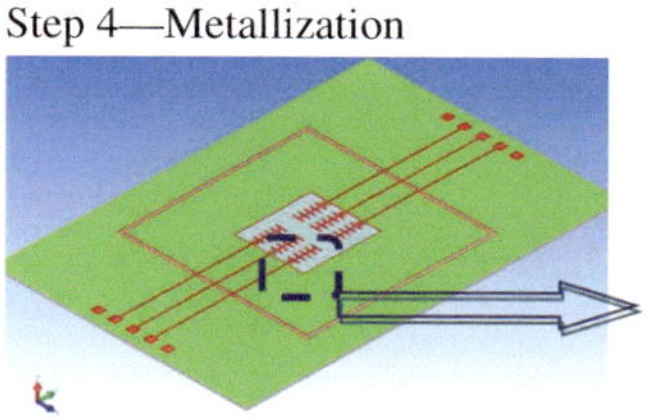

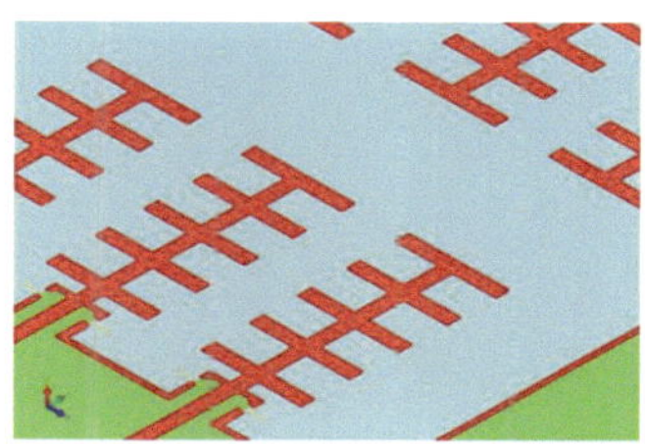

Step 5—DRIE sensor

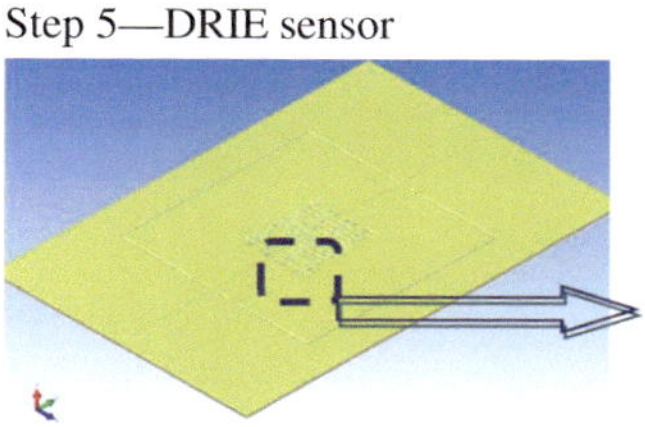

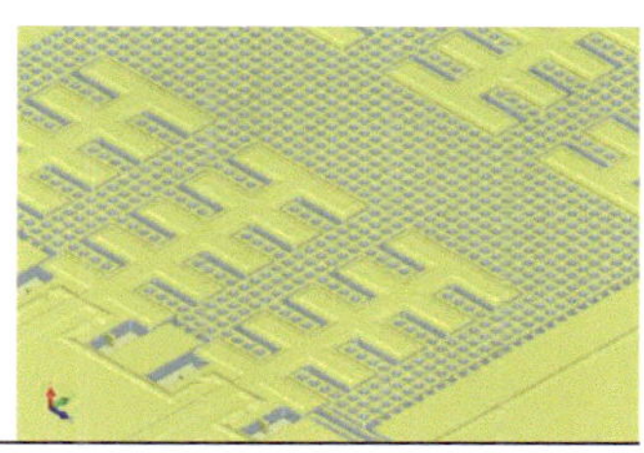

(continued)

Table 7.1 (continued)

Step 6—Sidewall passivation

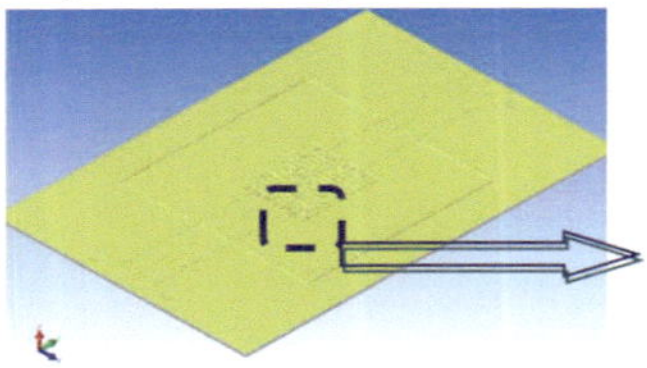

Step 7—Release etch

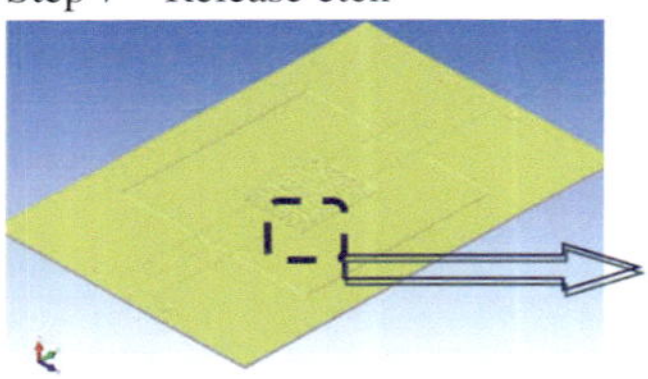

Step 8—Cover oxide removal

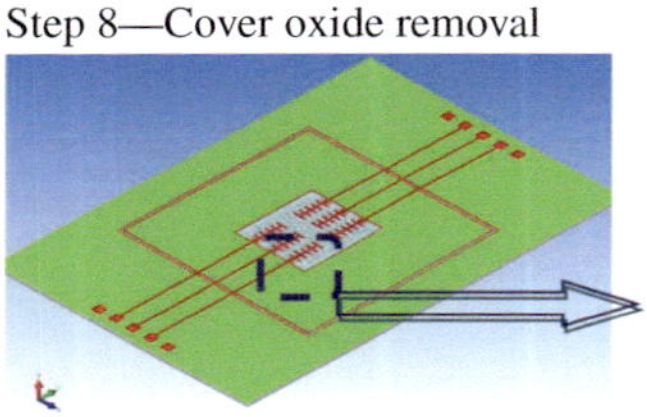

Step 9—Nitride etch

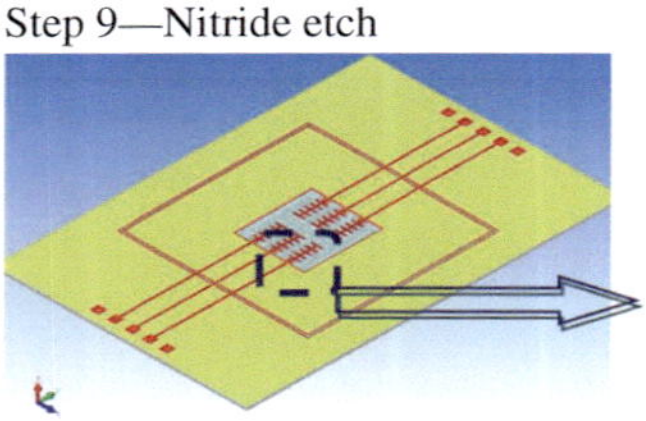

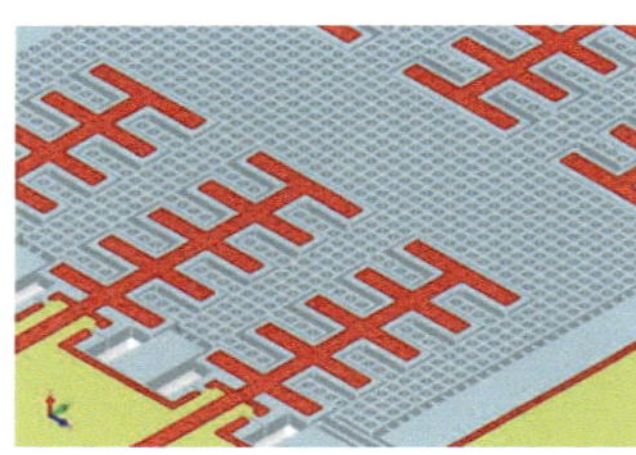

Step 10—Topcap wafer

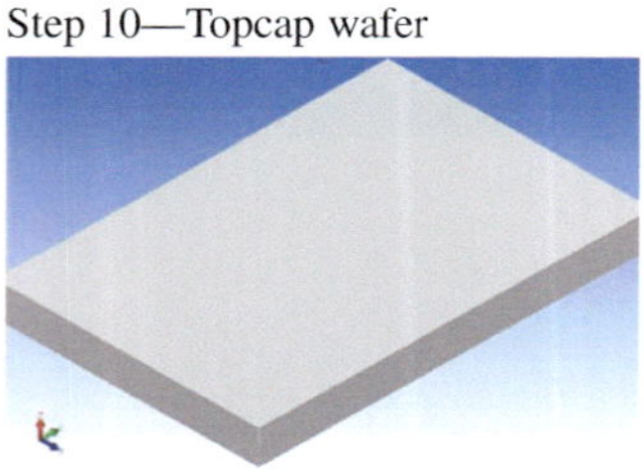

Step 11 – KOH etch

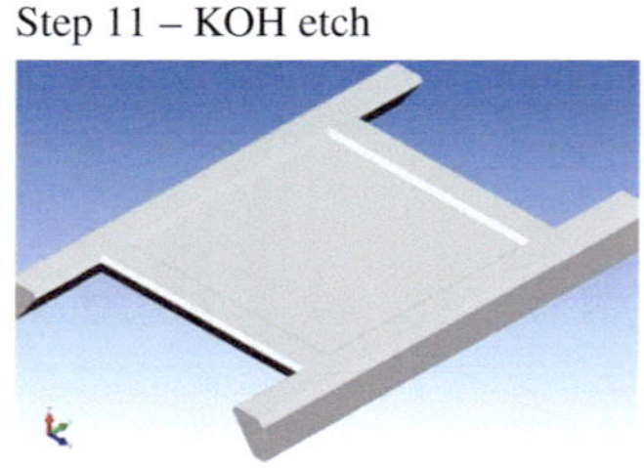

(continued)

Table 7.1 (continued)

Step 12—Topcap metal	Step 13 – Topcap oxide
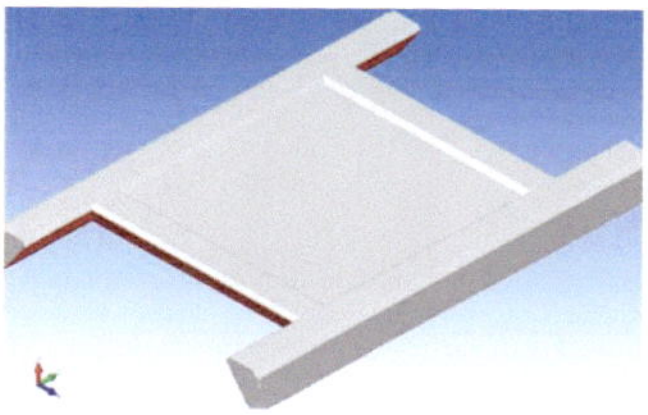	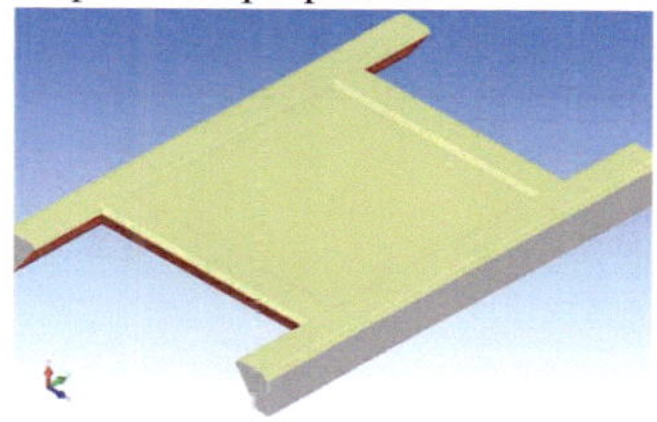
Step 14—Glassfrit print	Step 15 – Final wafer
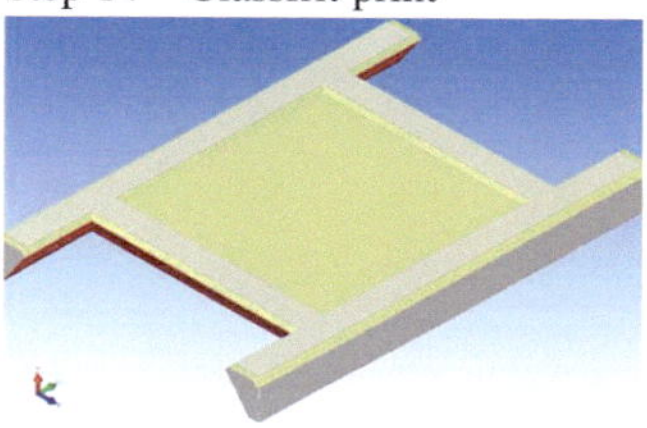	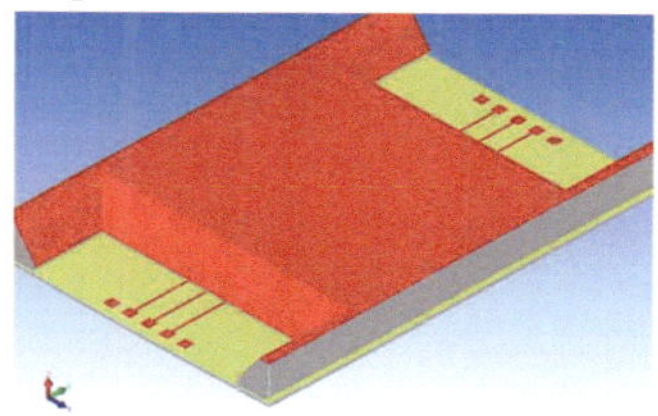

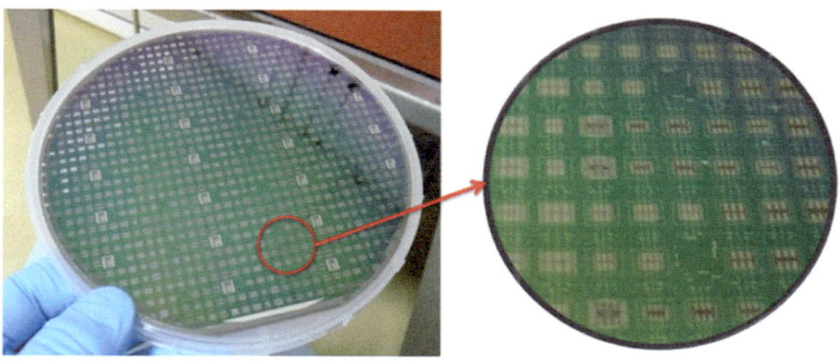

Fig. 7.12 Final not diced wafers with "naked" sensors for further investigation at ITE site [3]

7.3.1.3 Fabrication at X-FAB (1st Technological Run)

The fabrication process of ITE MEMS device at X-FAB took about 6–8 weeks. After fabrication of the structures ITE received chips from X-FAB (Figs. 7.12 and 7.13). Due to the fact that ITE and other X-FAB customer took separate runs, ITE has decided to split its twelve wafer lot into the following groups for the 1st technological run:

- 6 wafers with not capped sensors and not diced (as shown in Fig. 7.12);
- 6 wafers with the cap and with diced ready chips (as shown in Fig. 7.13)

The first group of wafers (not capped) was devoted for further investigation by ITE and its customer.

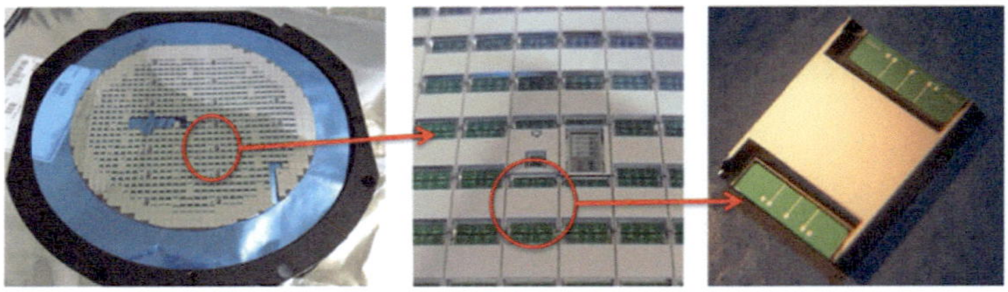

Fig. 7.13 Final diced wafers with capped chips ready to deliver to ITE customer [3]

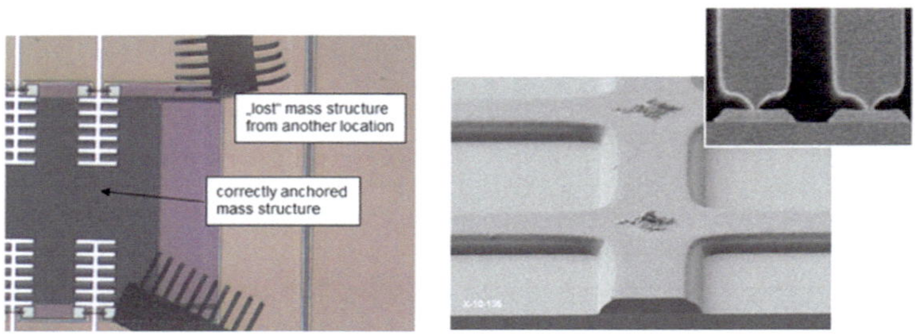

Fig. 7.14 Photos of issues of the 1st X-FAB run due to design and process issues

During the course of the project ITE was treated as a regular foundry customer by X-FAB. There has been no additional information flow and ITE did not receive any preliminary report about fabrication status, fabrication deviations, or problems before the X-FAB fabrication was completed. This was clearly a deficiency in communication because the 1st run did not yet fully adopt the tighter communication integration as proposed by the T^{2}M methodology. As a result several issues were found with the 1st run. Figure 7.14 shows a couple of sample photos indicating some of the issues as part of the X-FAB report. The left picture was taken with an optical microscope and shows some loose structures of the vibration mass. The middle inset shows a SEM picture with some residues while the right picture presents some not released parts due to a too short release etch time.

7.3.1.4 Preliminary Integration and Testing at ITE

Selected tasks were performed after the chips fabrication:

- Temporary integration of the sensor and the ASIC—(Fig. 7.15)
  The figure shows a few examples of different prototype packaging options only for initial tests. On the left side only the sensor was packaged. On the right side the sensor and the read-out electronics were connected and packaged.
- Characterization of selected sensors on the sensor level and system level and exemplary measurements results (Fig. 7.16). The goal of this measurement was to determine the resonance frequency of the sensor as reported in [16].

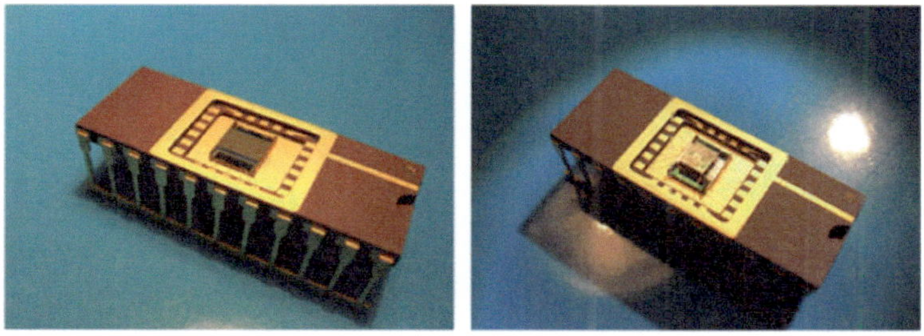

Fig. 7.15 Some examples of different prototype packaging options

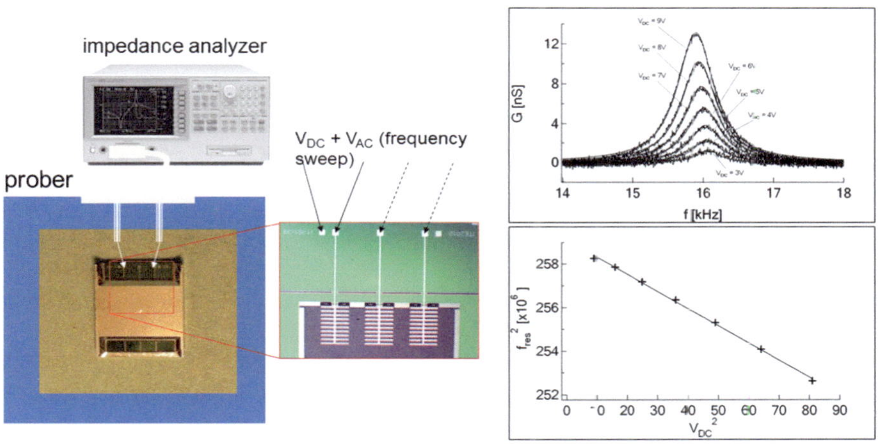

Fig. 7.16 Characterization of MEMS on wafer (sensor) and system level [5, 12, 16]

Similar integration and measurement tasks were performed partially next to the CORONA project at a site of the ITE customer.

7.3.1.5 Preparation of the 2nd Technological Run at X-FAB

The characterization work of the 1st run devices delivered valuable input for the 2nd technological run. After all the measurements were performed the best sensor architectures fulfilling the ITE customer's specification and requirements have been selected from the 18 prototypes fabricated during the 1st run. A new mask set with the same sensor structure but modified connection pads layout has been prepared for the 2nd run. The metallization level had to be redesigned. In the 2nd run the connection pads are located only on one side of the chip. Additionally the layout consists of two sensors placed together, one of them is rotated by 90 degrees. This solution made it easier to build a 2-axis (X and Y) sensor. In Fig. 7.17 one of

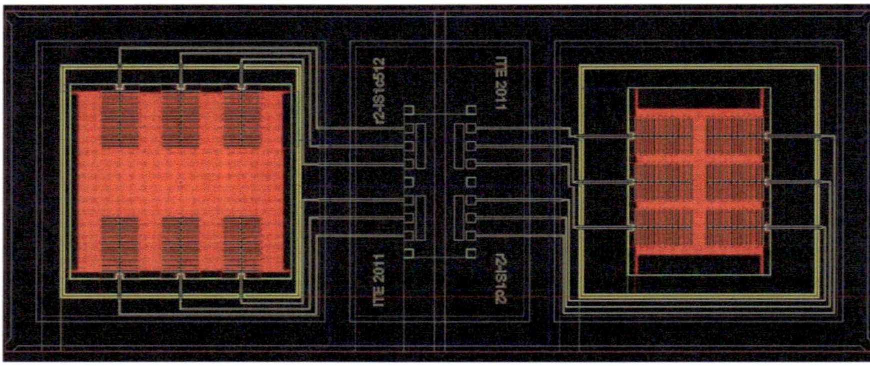

Fig. 7.17 Sample mask set for 2nd technological run at X-FAB

the four different architectures of the mask set is shown. For mask preparation the *CoventorWare* software was used as well *AutoCad*.

Before real fabrication at X-FAB similar virtual manufacturing work has been performed during the preparations of the 1st run. In Table 7.2 the sequence of virtual prototyping steps is shown for which the Coventor SEMulator3D was used. This virtual emulation consists only of the constructing of the sensor wafer, without silicon cover.

The 2nd fabrication run commenced without any problems. One additional complication of the 2nd technological run was that the ITE and another customer layout were mixed together on one mask set and fabricated on the same wafers. By this it was possible to further reduce the costs and to explore the fabrication of chips alongside with other companies ones. Figure 7.18 presents the ITE chips fabricated in the 2nd technological run.

Work concerning the design of a dedicated ASIC for read-out and calibration was ongoing in parallel. X-FAB XC-10 technology was preliminary selected at the first stage of the project. Due to financial constrains finally the ASIC was being designed in AMS C35 technology (by Austria Microsystems, 0.35 μm) available within Europractice MPW service (this was option evaluated during the initial project steps as one of the other options). The design of the dedicated ASIC and finally fabricated ASIC chips are presented in Fig. 7.19. Dedicated ASIC—optimized dual channel readout electronics module—main features:

- Two independent analog read-in channels,
- Single Oscillator,
- 2 Output Channels of Analog Data,
- 2 Output Channels of Digital Data (two SPI interfaces),
- Both Channels calibrated by external discrete elements—currently tested and optimized for capacitive MEMS accelerometers for 2G and 10G acceleration range.

Table 7.2 X-FAB virtual manufacturing run with Semulator 3D of ITEs design for 2nd run (sensor only)

Step 1—SOI wafer

Step 2—Etch and refill isolation trench

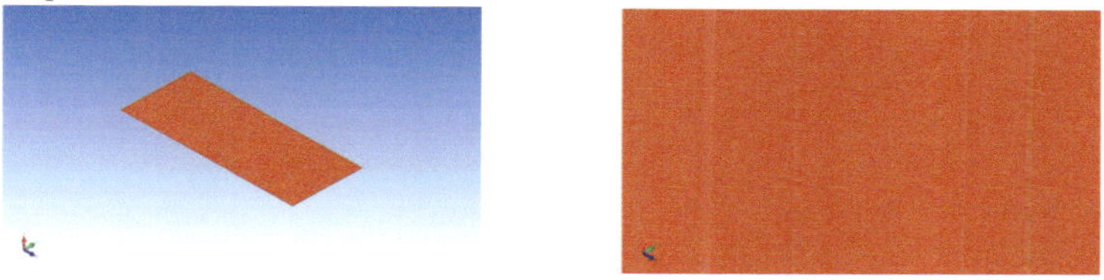

Step 3—Intermediate isolator

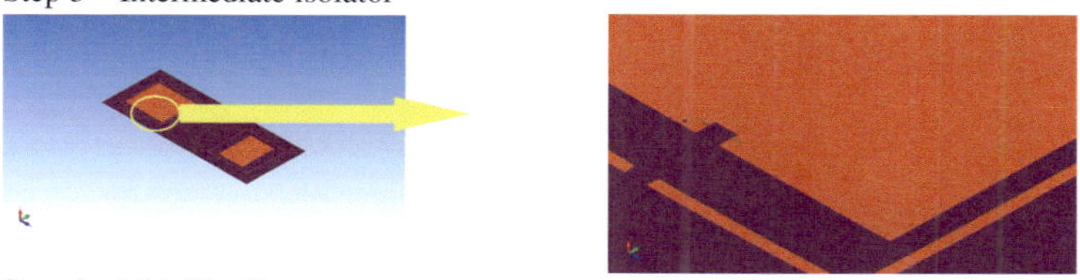

Step 4—Metallization

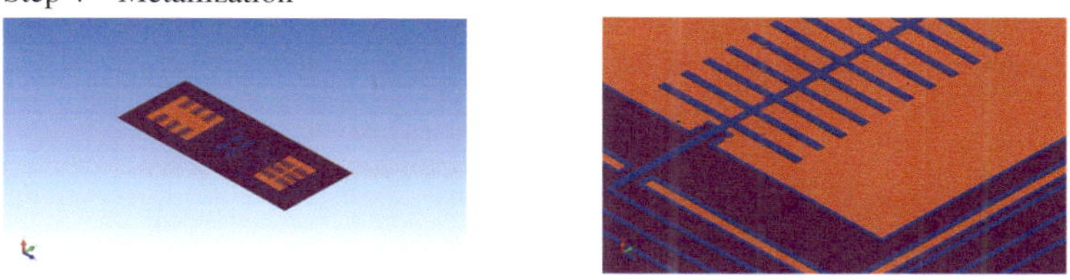

Step 5—DRIE sensor

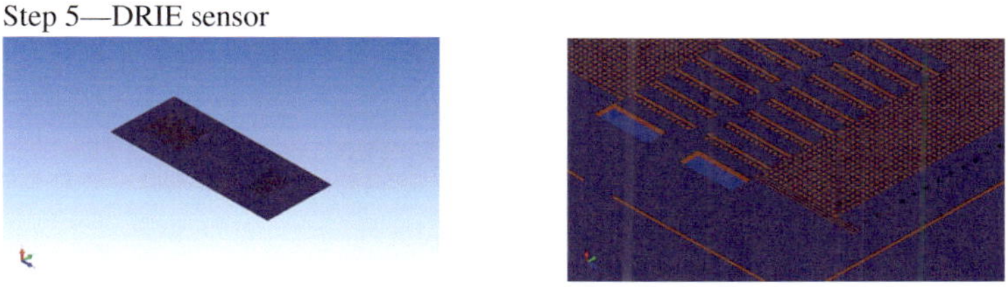

Step 6—Sidewall passivation

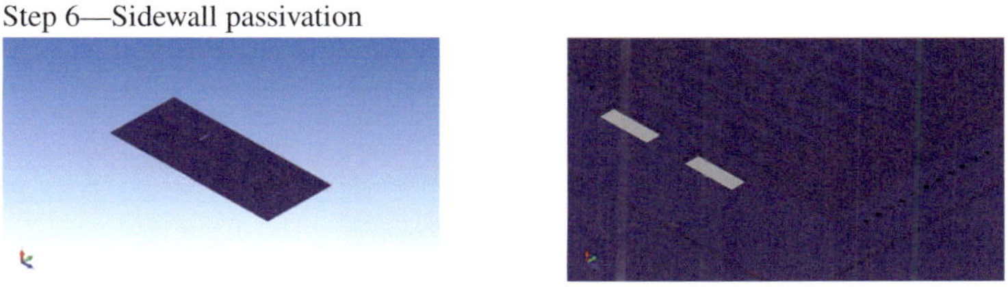

Step 7—Release etch

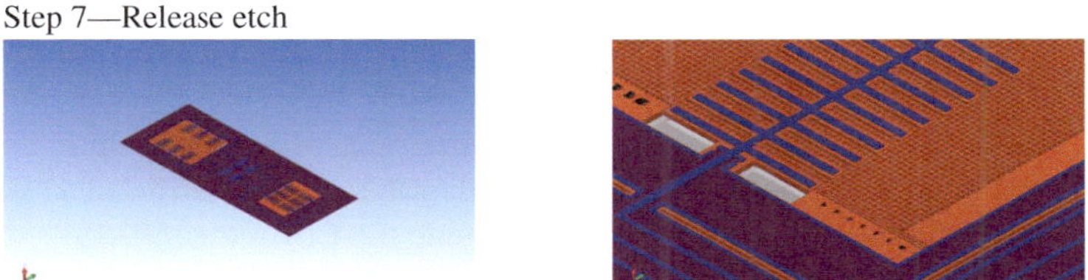

(continued)

Table 7.2 (continued)

Step 8—Cover oxide removal	Step 9—Nitride etch
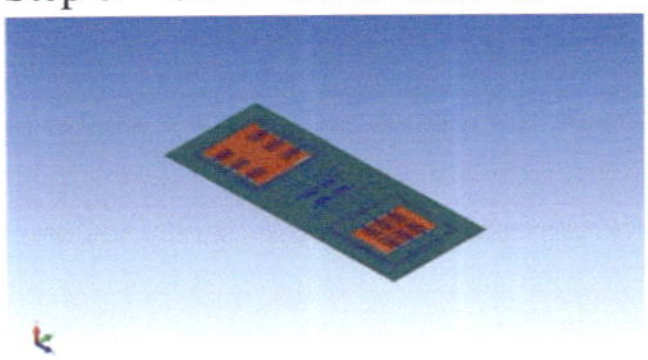	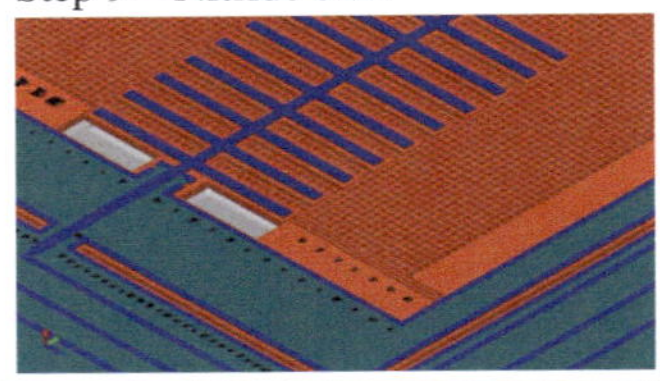

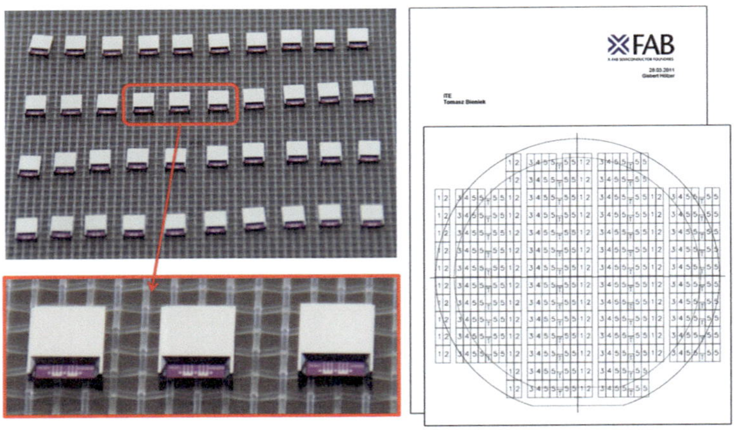

Fig. 7.18 ITE chips fabricated in the 2nd series and fabrication report

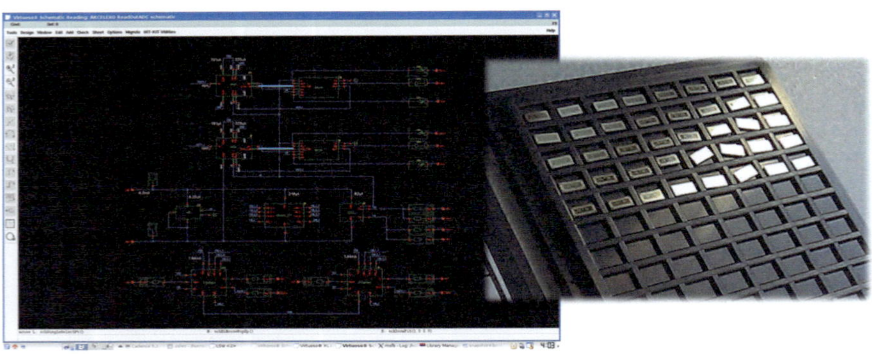

Fig. 7.19 Dedicated ASIC schematic and chips fabricated using Europractice MPW service

Finally the ITE customer successfully tested the designed system in a real automotive environment. Some work concerning the co-simulation of ITE sensor model with ASIC model has been performed using the Coventor MEMS+ software which is intended to perform specific co-simulations of electronics and mechanics (using an electronic model of the mechanics) within Cadence environment used

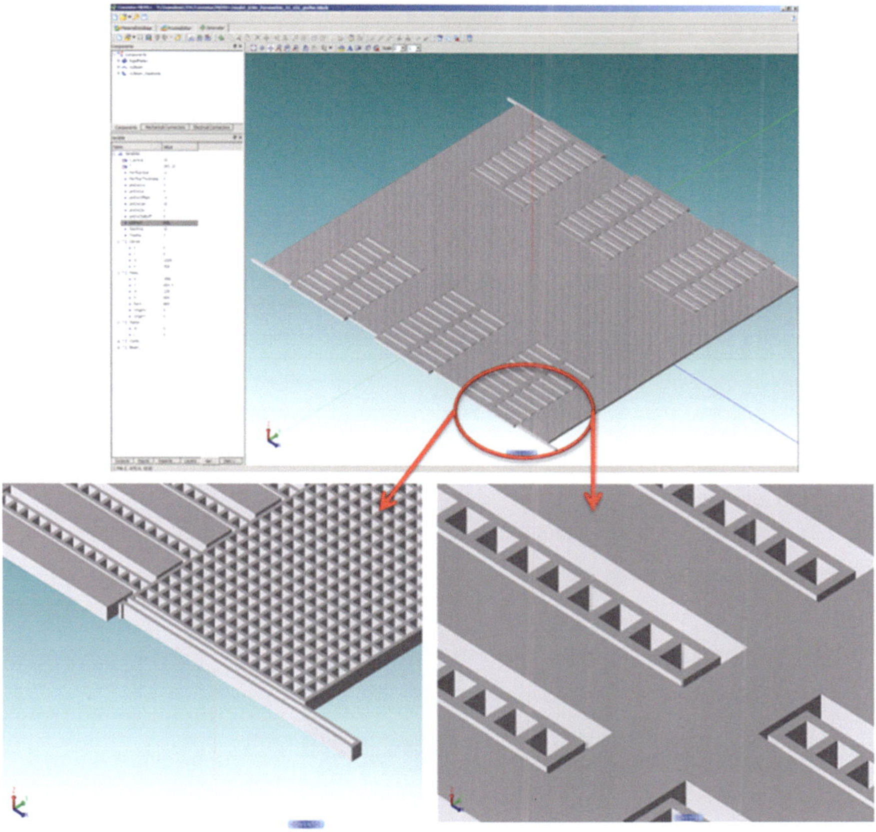

Fig. 7.20 Model of ITE sensor in Coventor MEMS+ environment

by ITE. It was a task of same importance as in case of MEMS sensor virtual validation performed before fabrication. The ASIC design had to be virtually tested and validated before expensive fabrication.

During the project period there have been some software implementation issues resulting that the alternative solution was build up. That alternative was that the model of the ITE sensor as shown in Fig. 7.20 to be simulated using a Verilog specification. Before the use of the MEMS+ software the corresponding simulation was performed using the ITE HEDORIS system. In Fig. 7.21 a few examples of the HEDORIS simulation of 1st series of ITE sensors are shown. After successful implementation of the thermal modules in HEDORIS, the system was developed to excess simulation range on mechanical simulation of the MEMS modules co-simulated with electronic modules inside the Cadence environment. It was used to speed up the design verification process.

ITE is not a commercial software developer. Therefore HEDORIS toolset cannot compete with commercial CAD tools. Nevertheless the current implementation confirmed its functionality and usefulness. Current works apart from formal

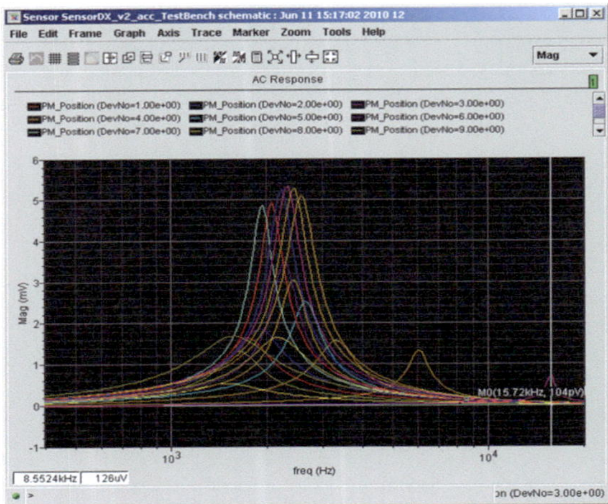

	Mass	Ncomb	Ks
r1-ISO.2e	2,02E-05	12	4,50E-01
r1-ISO.4d	3,24E-05	14	8,60E-01
r1-ISO.6d	2,87E-05	16	8,60E-01
r1-ISO.8d	2,41E-05	18	8,60E-01
r1.IS1a	1,83E-05	20	4,46E+01
r1.IS1b	1,83E-05	20	6,61E+00
r1.IS1c	1,83E-05	20	1,99E+00
r1.IS1c15	3,41E-05	20	1,99E+00
r1.IS1c27	2,85E-05	20	1,99E+00
r1.IS1c39	3,87E-05	20	1,99E+00
r1.IS1c51	5,83E-05	20	1,99E+00
r1.IS1d	1,83E-05	20	8,60E-01
r1.IS1e	1,83E-05	20	4,50E-01
r1-IS1.2c	4,65E-05	22	1,99E+00
r1-IS1.4c	4,03E-05	24	1,99E+00
r1-IS1.6c	3,71E-05	24	1,99E+00
r1-IS1.8c	3,37E-05	24	1,99E+00
r1-IS2c	3,02E-05	24	1,99E+00

Fig. 7.21 Table and graph view of all variants Verilog-AMS simulation results [5, 12]

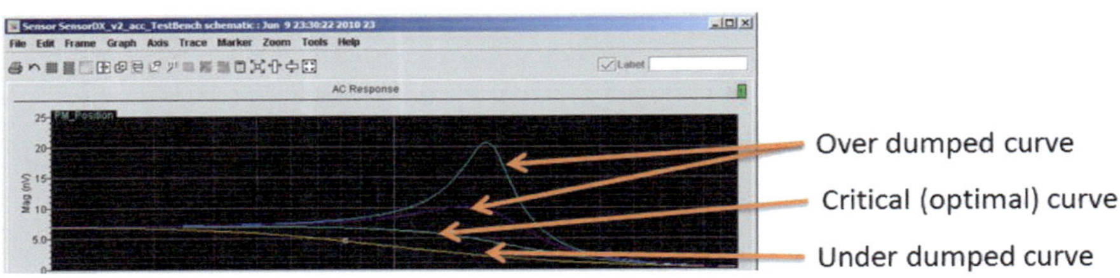

Fig. 7.22 Simulation results for four damping factors [5, 12]

verification of the MEMS model correctness cover bidirectional accuracy validation of developed HEDORIS model vs. MEMS+ model and finally vs. final devices integrated with just developed readout electronics module design. For the current product development a series of 18 variants of vibration sensor for 1st fabrication series were developed. After formal analysis, a Verilog-AMS model of the sensor was developed and included for excessive simulations covering various types of vibrations imposed to the structure assisted by several variants of readout electronic circuitry. Figure 7.21 presents a table of model parameters along with simulation results grabbed from Spectre. It shows data for all 18 variants of the structure used in the 1st technological run. The simulations with the HEDORIS system confirmed resonant frequencies coherent with expectations of the designers.

After detailed analysis of the simulation results, two structures have been selected for final development and integration to fit 1.9 kHz resonant frequency within given range of acceleration. One can find amplitude frequency characteristics for various damping factors to be calibrated for final device in Fig. 7.22. It was also compared: the designer prediction, HEDORIS simulation, and experimental measurements from the first fabrication run—Fig. 7.23.

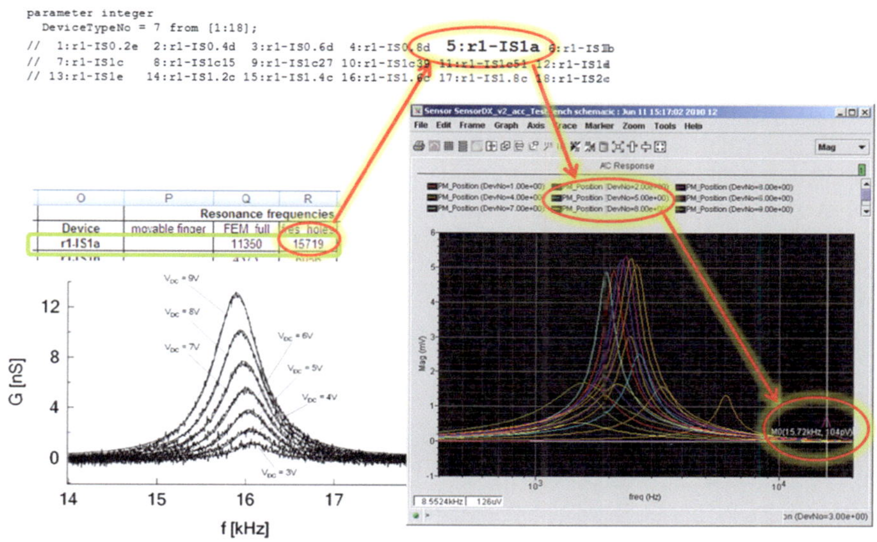

Fig. 7.23 Coherence between prediction, simulation, and measurement for particular MEMS structure

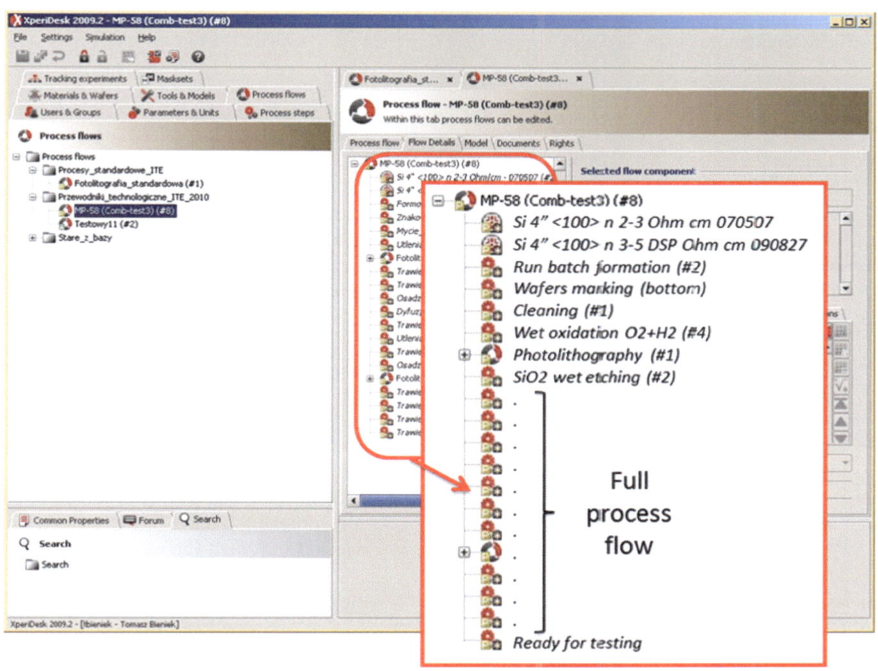

Fig. 7.24 XperiDesk system with one of the ITE CORONA-related test structures technological run with all technological processes

7.3.1.6 Process Development Execution System

Within the Prototype/Product Development Stage and within the project XperiDesk software from Process Relations has been tested as a Process Development Execution System (PDES) tool—Fig. 7.24. It was preliminary introduced for fabrication line management and optimization 'system. Full implementation of the XperiDesk toolset would be 'beneficial and looks promising, although due to the limited period of CORONA project time flow, such a solution was tested as feasibility study to verify needs, proposed solutions and potential benefits of such modernization of fabrication line management system.

7.4 Launch Phase

As briefly presented in Sect. 7.3, the business model of ITE as research institute rather stops with the successful demonstration of prototypes or demonstrator as well small volume/pilot production. Product market introduction usually is a part of ITE customer business process.

References

1. Bieniek, T., Janczyk, G., Grabiec, P., Szynka, J., Kalicinski, S., Janus, P.: Micro and nano device - customer-oriented product e-engineering in e-CoFrame. In: eChallenges e-2010 Conference Proceedings, Warszawa, 27–29 October 2010 (2010). IEEExplore, New York
2. Bieniek, T., Janczyk, G., Grabiec, P., Szynka, J., Kalicinski, S., Janus, P., Domanski, K., Sierakowski, A., Ekwinska, M., Szmigiel, D., Tomaszewski, D., Holzer, G., Schropfer, G.: Customer-oriented product engineering of micro and nano devices. In: Proceedings of the 17th International Conference on Mixed Design of Integrated Circuits and Systems (MIXDES), Wroclaw, Poland, 24–26 June 2010, pp. 81–84. IEEExplore, New York (2010)
3. Bieniek, T., Janczyk, G., Kalicinski, S., Janus, P., Domanski, K., Grabiec, P., Szynka, J.: Methodology and tools for customer-oriented product engineering of micro and nano devices. In: Proceedings of the 10th Conference "Electron Technology" ELTE 2010, CD, Wroclaw, Poland, 22–25 September 2010 (2010)
4. Bieniek, T., Janczyk, G., Janus, P., Szynka, J., Grabiec, P.: Determination of electrical and mechanical parameters in capacitive MEMS accelerometers using electrical measurements. In: Smart Systems Integration 2009, pp. 1192–1197. VDE Verlag GmbH, Berlin (2011)
5. Bieniek, T., Janczyk, G., Grabiec, P., Szynka, J., Kaliciński, S., Janus, P.: Methodology and tools for customer-oriented micro and nanodevices efficient product development. In: Proceedings of the 2nd International Conference Engineering and Meta-Engineering, Orlando, 27–30 March 2011, pp. 56–61 (2011)
6. COMSOL: Comsol – more info available at http://www.comsol.com (2013). Accessed 10 April 2013
7. Coventor, Inc.: Coventor Home Page. http://www.coventor.com (2013). Accessed 10 April 2013
8. EUROPRACTICE: Europractice service. http://www.europractice.com (2013). Accessed 10 April 2013

9. Holzer, G., Knechtel, R., Breit, S., Schropfer, G.: 3-D process modeling - a noveland efficient tool for MEMS foundry design support. In: Advanced Semiconductor Manufacturing Conference, ASMC '09, IEEE/SEMI, pp. 200–205 (2009)
10. ITE: InstytutTechnologiiElektronowej – more info available at http://www.ite.waw.pl (2013). Accessed 10 April 2013
11. Janczyk, G., Bieniek, T., Szynka, J., Grabiec, P.: Integrated thermal modeling of heterogeneous eCubes stacked devices. In: Proceedings of the Conference THERMINIC 2008 - 14th International Workshop on Thermal Investigations of ICs and Systems, Roma, Italy, 24–26 September 2008, pp. 80–84 (2008)
12. Janczyk, G., Bieniek, T., Szynka, J., Grabiec, P., Janus, P., Kalicinski, S.: Micro and nano device reliability control by MOS transistors mechanical stress sensitivity estimation and flexible, customer oriented product engineering flow. In: IEEE International Reliability Workshop (IIRW10), South Lake Tahoe 17–21 October 2010 (2010)
13. Janczyk, G., Bieniek, T., Grabiec, P., Wymyslowski, A., Swierczynski, R.: Optimization on MEMS-IC SiP development reliable design methods. In: Proceedings of the International Workshop on Testing Three-Dimensional Stacked Integrated Circuits 3D-TEST in Conjunction with ITC, Anaheim, 4–11 November 2012 (2012)
14. Kaliciński, S., Bieniek, T., Janus, P., Grabiec, P.: Reliability issues in modeling and simulations of the heterogeneous integrated systems. Microelectron. Reliabil. **69**, 428–431 (2009)
15. Kalicinski, S., Janus, P., Bieniek, T., Domanski, K., Ekwinska, M., Sierakowski, A., Szmigiel, D., Grabiec, P.: Rapid prototyping of electrostatically-driven MEMS. Elektronika LI(6), 114–115 (2010)
16. Kalicinski, S., Bieniek, T., Janus, P., Grabiec, P.: Determination of electrical and mechanical parameters in capacitive MEMS accelerometers using electrical measurements. Microelectron. Reliab. **51**, 1192–1197 (2011)
17. Process Relations GmbH, XperiDesk – more info available at, http://www.process-relations.com (2012). Accessed 10 April 2013
18. X-FAB: More info available at www.xfab.com (2013). Accessed 10 April 2013

Chapter 8
Details for Managing a Project

This chapter gives more details about the project management sequences, the steps of the business processes and their inputs and outputs. The sequences and steps are basically taken from the PRINCE2® standard, partly extended or slightly rearranged. As an overview the full PRINCE2® process map is presented in Fig. 8.1. Parts of this figure are extracted in the following descriptions for better readability. The inputs and outputs are taken from PRINCE2® as well, while a full set of templates for all project management and product engineering artefacts has been created. This set contains all templates advised by the PRINCE2® standard with additions of further predefined deliverables. This set is contained within the default *Electronic Project Binder*. The set of templates and the *Electronic Project Binder* is defined in more detail in Sect. 6.5.

8.1 The Management of the Project Preparations

The project preparations are performed in the *Project Initialization Stage* of the engineering method (Sect. 5.4). In detail the Stage performs the steps summarized in Table 8.1. The table shows the inputs and outputs of the activities. They are graphically presented with their full context in Figs. 8.1 and 8.2 as an excerpt for better readability. An item listed as input and output becomes updated during the activities of this Stage. Besides external inputs or outputs, all inputs and outputs are retrieved and stored from and to the *Electronic Project Binder* (see below) referenced as "Management Information" in the drawings. The activities are executed in the sequence of the table rows and in accordance with the PRINCE2® best practices (e.g., [1,4]). The flow of business process steps from an abstract point of view can be seen in Fig. 5.9 as well. The abbreviations SU, IP, DP, PL, CS, MP, SB, and CP followed by a number refer to the corresponding PRINCE2® process groups and the step number. They stand for SU = Starting Up a project, IP = Initiating a Project, DP = Directing a Project, PL = Planning, CS = Controlling a Stage, MP = Managing Product delivery, SB = Managing Stage Boundaries, and CP = Closing a Project.

D. Ortloff et al., *MEMS Product Engineering*, DOI 10.1007/978-3-7091-0706-5_8,

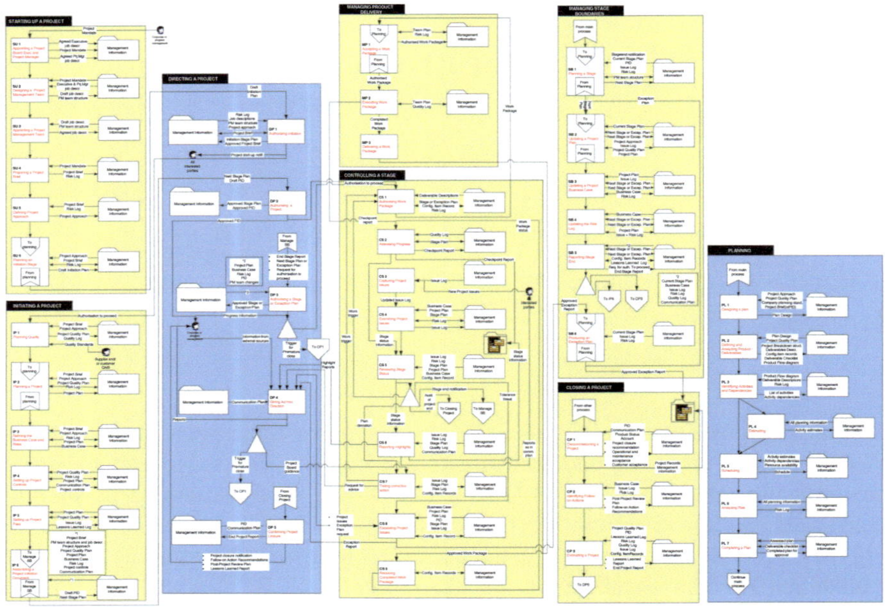

Fig. 8.1 Complete PRINCE2® process map (as an overview, excerpts follow in separate graphics)

8.1.1 *Process Group Planning*

Some sequences of process steps can be considered as modules which are plugged into or used by process steps of other groups. One of these generic activity sequences are the planning steps which are used to generate the overall Project Plan, derive Stage plans, plan a Work Package, etc. A general introduction into the planning efforts and the importance of the planning and tracking is given in Sect. 8.4.1. The steps of a planning subprocess are presented in Table 8.2. The table shows the different steps to create a plan showing the required inputs and outputs. A graphical representation of the process sequence is given in Fig. 8.3. More details about the planning activities and its best practices and how a good plan can be created are given in Sect. 8.4.1 and in, e.g., [1, 4].

For the planning it is important that the deliverables-based planning paradigm of PRINCE2® is applied and that tasks generated to create these deliverables are following the T^2M to achieve customer information, partner coordination, and so forth. In summary this means that not the activities are the start for the planning efforts, the deliverables (products in PRINCE2® terms) for the next Gate are the starting point. These deliverables and their development are the highest level in the project plan and are successively broken down into work packages, activities

Table 8.1 Project initialization stage activities with inputs and outputs

Inputs	Activities	Outputs
Project Mandate	SU1 Appointing a Project Board Executive and Project Manager Set up the electronic project binder	Agreed Executive job description Agreed Project Manager job description Electronic project binder
Agreed Executive and Project Manager job description Project Mandate	SU2 Designing a Project Management Team	Draft job description PM team structure
Draft job descritpion PM team structure	SU3 Appointing a Project Management Team	Agreed job descriptions
Project Mandate	SU4 Preparing a Project Brief	Project Brief Risk Log
Project Brief Risk Log	SU5 Defining Project Approach	Project Approach
Project Approach Project Brief Risk Log	SU6 Planning an Initiation Stage(using the process sequence in table 8.2 as subprocess)	Draft Initiation Plan Risk Log
Draft Initiation Plan Risk Log Job descriptions PM team structure Project approach Project Brief	DP1 Authorising Initiation	Initiation Stage Plan Approved Project Brief Project Brief
Authorisation to proceed Project Brief Project Approach Quality Standards from Supplier and/or customer QMS	IP1 Planning Quality	Project Quality Plan Quality Log
Project Brief Project Approach Project Quality Plan Risk Log	IP2 Planning a Project(using the process sequence in table 8.2 as subprocess)	Project Plan Risk Log
Project Brief Project Approach Risk Log Project Plan	IP3 Refining the Business Case and Risks	Business Case Risk Log Project Plan
Project Quality Plan Risk Log Project Plan	IP4 Setting up Project Controls	Communication Plan Project controls Risk Log Project Plan
Project Plan Project Quality Plan	IP5 Setting up Project Files	Issue Log Lessons Learned Log Project Quality Plan
Project Brief PM team structure and job descr. Project Approach Project Quality Plan Project Plan Business Case Risk Log Project controls Communication Plan	IP6 Assembling a Project Initiation Document including planning the next Stage(using the process sequence in table 8.4 as subprocess)	Draft PID Next Stage Plan

for the work packages and tasks. To achieve this breakdown initially a *Product Breakdown Structure* (PBS) is created. This PBS recursively splits down the end product into smaller deliverables (divide-and-conquer strategy). Each deliverable (on all levels) needs to be documented using the Deliverable Description template (PRINCE2®) building the contract between the different involved parties. These

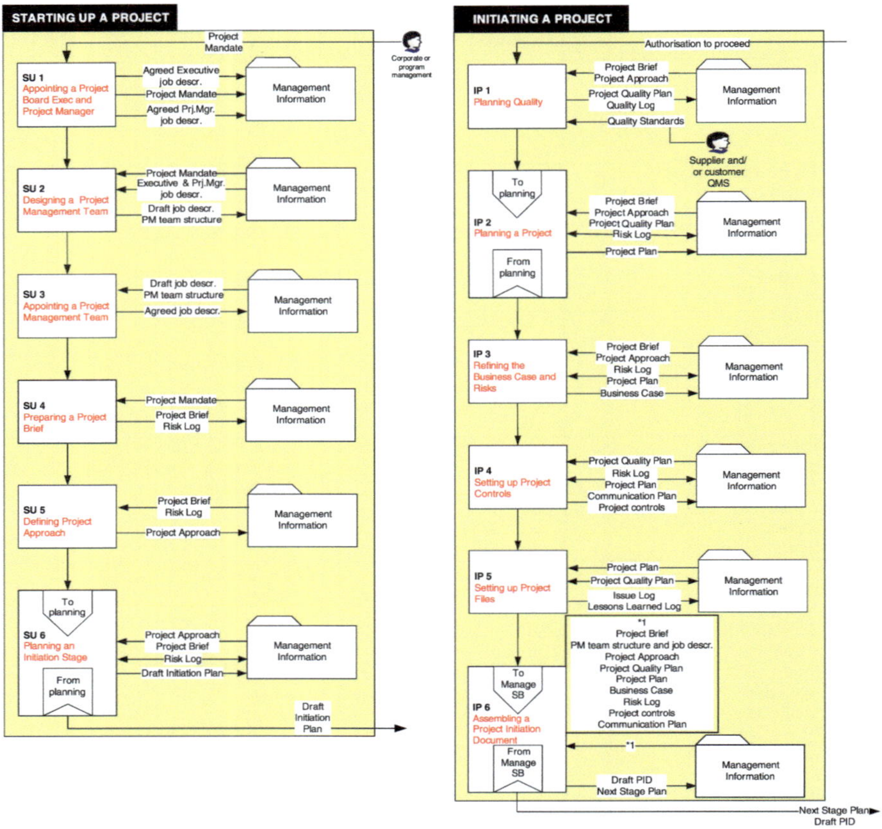

Fig. 8.2 PRINCE2® processes starting up a project and initiating a project

descriptions build the basis for creating the Work Package Descriptions (template) used to execute the engineering efforts. From the deliverables list the Product Flow Diagram (PFD), a dependency graph, can be created. This graph enables a proper scheduling and resource assignment via a project planning tool, e.g., MS Project. More details about the product/deliverables-based planning approach can be found in [1, 4].

8.2 The Management of the R&D Stages

Like the management of the other method aspects, the management of the different R&D Stages of the *Startup Phase* and the *Development Phase* (according to Fig. 5.3) is performed according to the PRINCE2® project management standard. Because of

Table 8.2 Planning activities with inputs and outputs; used in several places

Inputs	Activities	Outputs
Project Approach Project Quality Plan Company planning standards Project Brief (or PID)	PL1 Designing a plan	Plan Design
Plan Design Project Quality Plan	PL2 Defining and Analyzing Product/Deliverables	Project Breakdown Structure(PBS) Deliverable Descriptions Configuration item records Product/Deliverables Checklist Product Flow diagram
Product Flow diagram Deliverables Descriptions Risk Log	PL3 Identifying Activities and Dependencies	List of activities Activity dependencies
All planning information	PL4 Estimating (optional and only used where applicable)	Activity estimates
Activity estimates Activity dependencies Resource availability	PL5 Scheduling	Schedule
All planning information	PL6 Analyzing Risk	Updated Risk Log
Assessed plan Product/Deliverable checklist	PL7 Completing a Plan	Product/Deliverable checklist Completed plan for approval

that the following descriptions are rather brief to not replicate PRINCE2® literature like [1,4]. To guide the project managers and Product Engineering efforts the overall flow and sequence of process steps is presented. More information about the details of the different process groups, process steps, and activities can be extracted from literature.

The management process steps can be grouped into certain process groups like Planning Activities, Managing Stage Boundaries, etc. This grouping is already included in the PRINCE2® standard and can be seen in Fig. 8.5 together with Fig. 8.6.

Apart from this standard grouping into the PRINCE2® standard process groups a more flow-oriented grouping (Preparing a Project, Planning in various forms and steps, Inside a Stage Loop, Finalizing a Stage, Exception Treatment, Gate Assessment and Finishing a Project) can be set up. This structure is used when describing and depicting the serialized PRINCE2® process of this method. The bottom part of it can be seen in Fig. 8.4. The following sections will describe these flow-oriented process groups in a bit more detail.

8.2.1 Process Group Inside a R&D Stage Loop

The main engineering activities for a project (technical, marketing and sales, financial, legal, etc.) are managed by the R&D Stage loop highlighted with the yellow box in Fig. 8.4. Here only the managerial steps and artefacts (in parts; all artefacts according to PRINCE2® are shown in Figs. 8.5 and 8.6 and listed in Table 8.3) are highlighted, the technical, financial, etc. deliverables are described within the individual Stages and Gates.

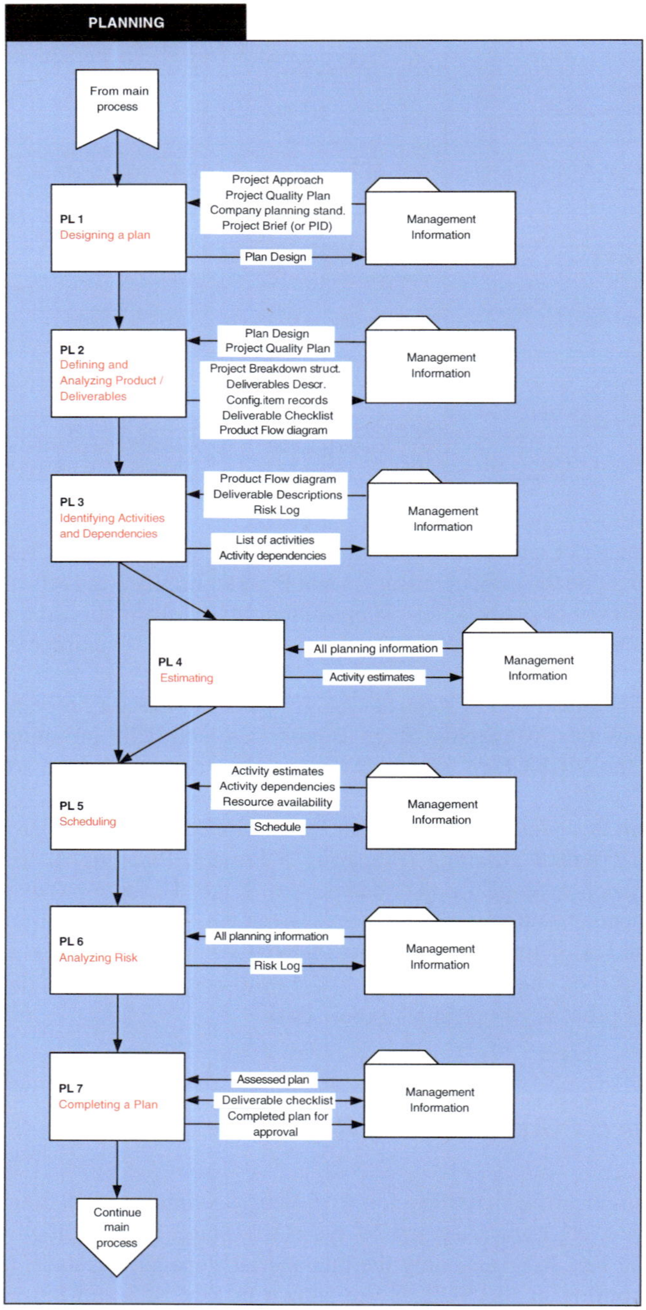

Fig. 8.3 Planning process sequence

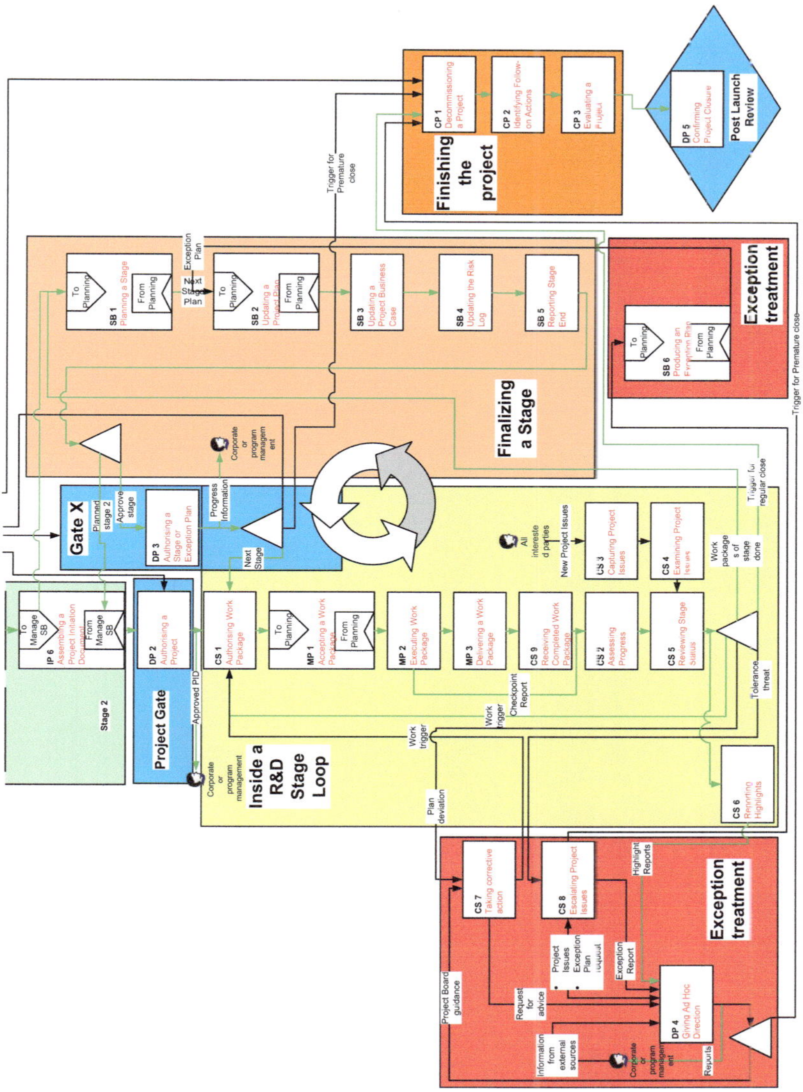

Fig. 8.4 Serialized method bottom part

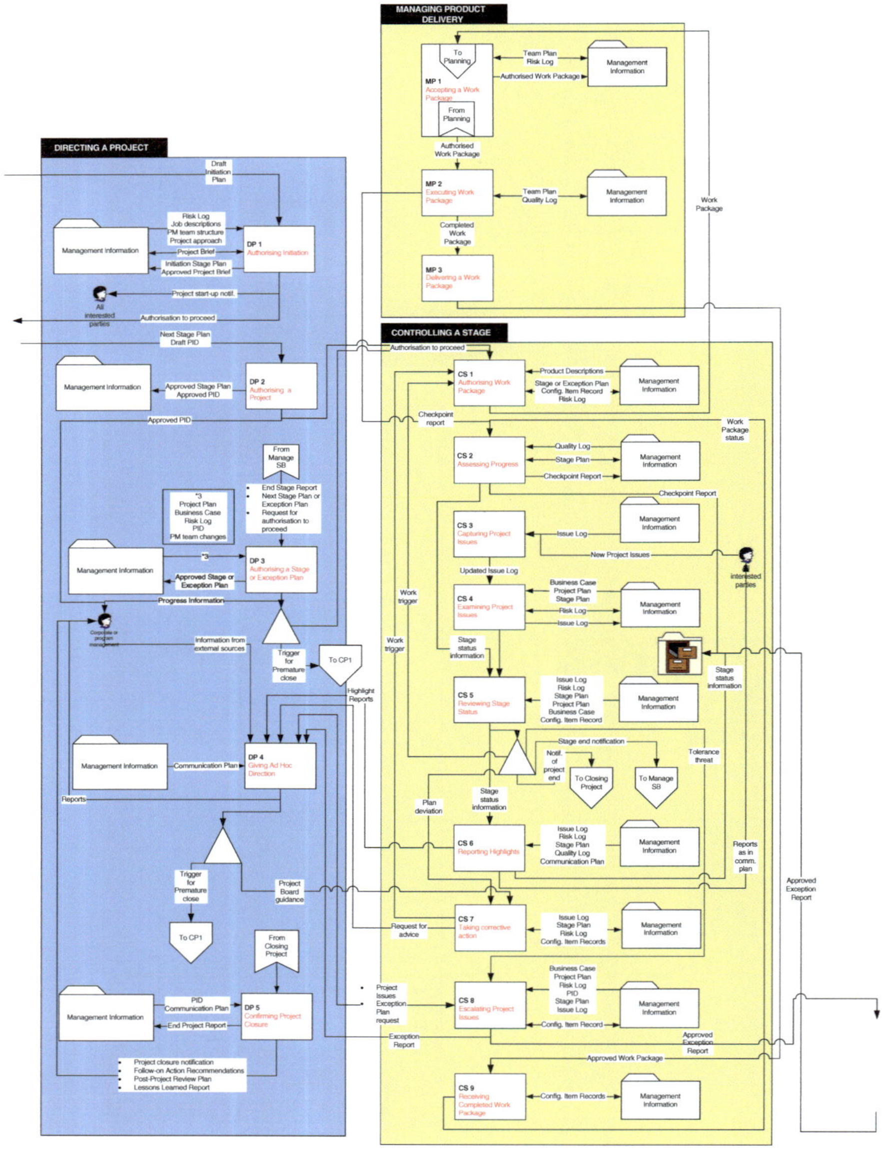

Fig. 8.5 PRINCE2® process directing a project, managing product delivery, and controlling a stage

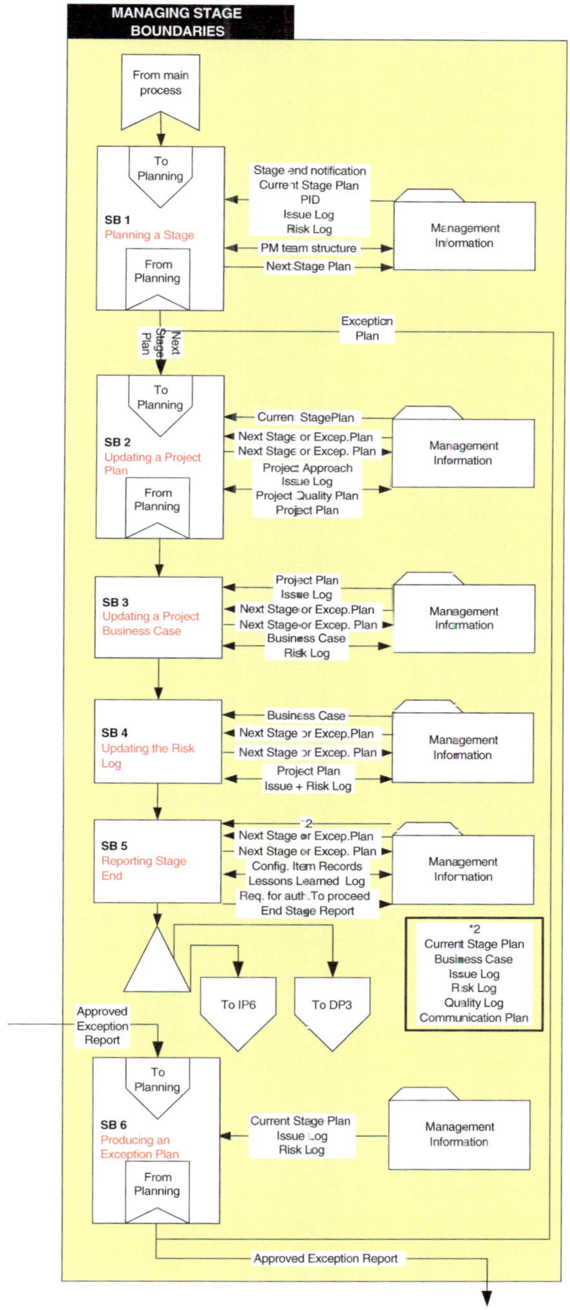

Fig. 8.6 PRINCE2® process managing stage boundaries

Table 8.3 Inside an R&D loop activities with inputs and outputs

Inputs	Activities	Outputs
Authorisation to proceed Approved Stage or Exception Plan Deliverable Descriptions Configuration Item Record Risk Log	CS1 Authorising Work Package	Approved Stage or Exception Plan Configuration Item Record Risk Log
Team Plan Risk Log	MP1 Accepting a Work Package	Updated Team Plan Updated Risk Log Authorised Work Package
Authorised Work Package	MP2 Executing Work Package	Completed Work Package Checkpoint report Updated Team Plan Updated Risk Log
Completed Work Package	MP3 Delivering a Work Package	Approved Work Package
Approved Work Package Configuration Item Records	CS9 Receiving Completed Work Package	Configuration Item Records Work Package status
Work Package status Checkpoint Report Quality Log Stage Plan	CS2 Assessing Progress	Stage Plan Checkpoint Report (to Management Information and Project Archive) Stage status information
Issue Log New Project Issues	CS3 Capturing Project Issues (optional only if triggered by new issues from log or customer)	Updated Issue Log
Business Case Project Plan Stage Plan Risk Log	CS4 Examining Project Issues(optional only if triggered by new issues from log or customer)	Updated Risk Log Updated Issue Log
Stage status information Issue Log Risk Log Stage Plan Project Plan Business Case Configuration Item Record	CS5 Reviewing Stage Status	(no direct output only decision where to continue as below)
Stage status information Issue Log Risk Log Stage Plan Quality Log Communication Plan	CS6 Reporting Highlights	Highlight Report Stage status information(to archive)
depending on the Stage status either of:		
	Stage not finished: trigger CS1 in this table to continue work	
	Stage finished: trigger to Finalizing a Stage in table 8.4	
	plan deviation detected: continue with CS7 exception treatment table 8.6	
	tolerance threat detected: continue with CS8 in table 8.6	
	all projects work done: trigger to Close Project in table 8.7	

The process steps of this group authorize the different work packages and manage the execution of the work package (including planning the work package). The work package planning, in contrast to the standard PRINCE2®, uses the T^2M methodology to plan the different Tasks and Transitions for efficient coordination between the collaborating partners. The steps of the project management process have to handle internal and external issues or exceptions and keep track of the project status and progress and report about it. Necessary corrective actions are initiated and their progress is coordinated with the projects executive. More details about the input and output of the different processes are presented in Table 8.3. The table also gives more details about the conditional executions of this loop. Figure 8.5 presents the parts of the PRINCE2® standard responsible for driving the work.

8.2.2 Process Group Finalizing a Stage

After all Work Packages and Deliverables of a Stage have been successfully completed, the process steps of the Finalizing a Stage group are executed. The steps of this subprocess basically round up the current Stage and prepare the next Stage. For that a plan for the next Stage is produced and the overall Project Plan, the Business Case and the Issue and Risk Logs become updated. Finally a Stage End Report triggering the next Gate assessment is generated. More details about the input and output of the different processes are presented in Table 8.4.

8.2.3 Process Group Gate Assessment

This process group, actually only one process step and decision making, is the key for successful product engineering. As outlined in the description of the product engineering method, e.g., in Sect. 8.4.1.2, the Gates are the essential check points during the lifetime of a project. They are essential to distinguish between the winning and failing project ideas. The importance and procedure of the Gate assessments is described in Sect. 8.4.1.2 and should not be re-discussed here. Only the managerial inputs and outputs are listed in Table 8.5.

8.2.4 Process Group Exception Treatment

In the process group Exception treatment the Project Steering Committee becomes actively involved in the project. This committee consists of the Project Executive, the Project Manager, and the Corporate Program Management. Usually the project is managed only by the Project Manager and Projects Executive. In the exception case, the Steering Committee gets the role to give project guidance, settle project issues, and approve Exception Plans. On the other hand the Steering Committee might feed information from external sources into the project, potentially changing the course of the project. In extreme cases the project might be even triggered for a premature end. More details of the process steps to treat exceptions are presented

Table 8.4 Finalizing a stage activities with inputs and outputs; used in several places

Inputs	Activities	Outputs
Stage end notification Current Stage Plan Project Initiation Document(PID) Issue Log Risk Log PM team structure	SB1 Planning a Stage (using the process sequence in table 8.2)	PM team structure Next Stage Plan
Current Stage Plan Next Stage or Exception Plan Project Approach Issue Log Project Quality Plan Project Plan	SB2 Updating a Project Plan (using the process sequence in table 8.2)	Next Stage or Exception Plan Project Approach Issue Log Project Quality Plan Project Plan
Project Plan Issue Log Next Stage or Exception Plan Business Case Risk Log	SB3 Updating a Project Business Case	Next Stage or Exception Plan Business Case Risk Log
Business Case Next Stage or Exception Plan Project Plan Issue + Risk Log	SB4 Updating the Risk Log	Next Stage or Exception Plan Project Plan Issue + Risk Log
Current Stage Plan Business Case Issue Log Risk Log Quality Log Communication Plan Next Stage or Exception Plan Config. Item Records Lessons Learned Log	SB5 Reporting Stage End	Next Stage or Exception Plan Config. Item Records Lessons Learned Log Request for authorisation to proceed End Stage Report
depending on context either:		
	continue IP6	
	trigger DP3 in table 8.5	

Table 8.5 Performing gate assessment with inputs and outputs

Inputs	Activities	Outputs
End Stage Report Next Stage Plan or Exception Plan Request for authorisation to proceed Project Plan Business Case Risk Log PID PM team changes	DP3 Authorising a Stage or Exception Plan	Approved Stage or Exception Plan Progress Information(to corporate or program management)
depending on decision either:		
	authorization to proceed: trigger CS1 in table 8.3	
	Trigger to premature end of project continuing with Close Project in table 8.7	

in Table 8.6 below, together with the inputs and outputs of the steps. The steps are labeled with exception treatment in Fig. 8.4 and belong to different process groups in the PRINCE2® standard (see step names). In the description here they are grouped together into this new group to mark them as being not necessarily part of a standard project life cycle.

Table 8.6 Exception treatment activities with inputs and outputs

Inputs	Activities	Outputs
Issue Log Stage Plan Risk Log Configuration Item Records	CS7 Taking corrective action	Issue Log Stage Plan Risk Log Configuration Item Records
depending on whether necessary or already given:		
	advice required: trigger DP4 in table 8.6	
	no advice required or already given: trigger CS1 in table 8.3 to continue work	
Communication Plan Information from external sources Highlight Report Request for advice Exception Report Project Issues Exception Plan request	DP4 Giving Ad Hoc Direction	Project Issues Exception Plan request Reports(to corporate or program management)
either:	Trigger to premature end of project continuing with Close Project in table 8.7	
	Project Board guidance continuing with CS7 in this table	
Tolerance threat Project Issues Exception Plan request Business Case Project Plan Risk Log PID Stage Plan Issue Log	CS8 Escalating Project Issues	Configuration Item Record Exception Report Approved Exception Report
depending on whether already escalated:		
	unsolvable issues or exception report: trigger DP4 in table 8.6	
	exception report approved or exception plan requested: trigger SB6 in table 8.6	
Approved Exception Report Current Stage Plan Issue Log Risk Log	SB6 Producing an Exception Plan(using the process sequence in table 8.2)	Exception Plan
	continue with SB2 in table 8.4	

8.3 Management of Finishing the Project

The process steps of managing the Finishing of the project are responsible to cleanly close a project and release the resources, including the personal working on the project. For that it uses the PRINCE2® processes consisting of decommissioning the Project Team, identifying and documenting the project follow-up actions and to evaluate the project. The findings are summarized in a Lessons Learned Report and an End Project Report. More details about the input and output of the different processes are presented in Table 8.7. A graphical representation of the PRINCE2® standard of these steps is presented in Fig. 8.7.

Table 8.7 Closing a project activities with inputs and outputs

Inputs	Activities	Outputs
PID Communication Plan Product Status Account	CP1 Decommissioning a Project	Project closure recommendation Operational and maintenance acceptance Customer acceptance Project Records(to archive) Management Information(to archive)
Business Case Issue Log Risk Log	CP2 Identifying Follow-on Actions	Post-Project Review Plan Follow-on Action Recommendations
Project Quality Plan PID Lessons Learned Log Risk Log Quality Log Issue Log Configuration Item Records	CP3 Evaluating a Project	Lessons Learned Report End Project Report

8.4 Project & Product Management Techniques

8.4.1 Planning

8.4.1.1 Features Provided/Addressed

For the project coordination and management the consequent Project Planning and Tracking is essential. These efforts address the features:

- Target focused, by awareness of problems and deviations,
- Control, by having the status easily accessible and
- Overall coordination, by enabling project management.

The importance of Planning and Tracking is emphasized by the Product Engineering Method as well. Here the planning processes are required by several other processes of the method.

8.4.1.2 Definition

Project Planning and Tracking is the central enabler of Project Management. A good introduction into the subject of Project Planning is given in [3]. Without constantly Planning and Tracking the project progress, including necessary replanning and rescheduling, it is impossible to have control. Therefore these efforts are the key to enable control and need to maintain a set of diverse plans (e.g., the *Project Plan*, *Stage Plans*) and regularly update the completion status. Having control is essential to be able to detect deviations and define and execute contingency plans. These can ensure that project goals are reached although unforeseen events were encountered. All planning and tracking efforts need to acknowledge project planning best practices as described in, e.g., [2, 3]. The later one gives a good comparison about different project management approaches and methodologies.

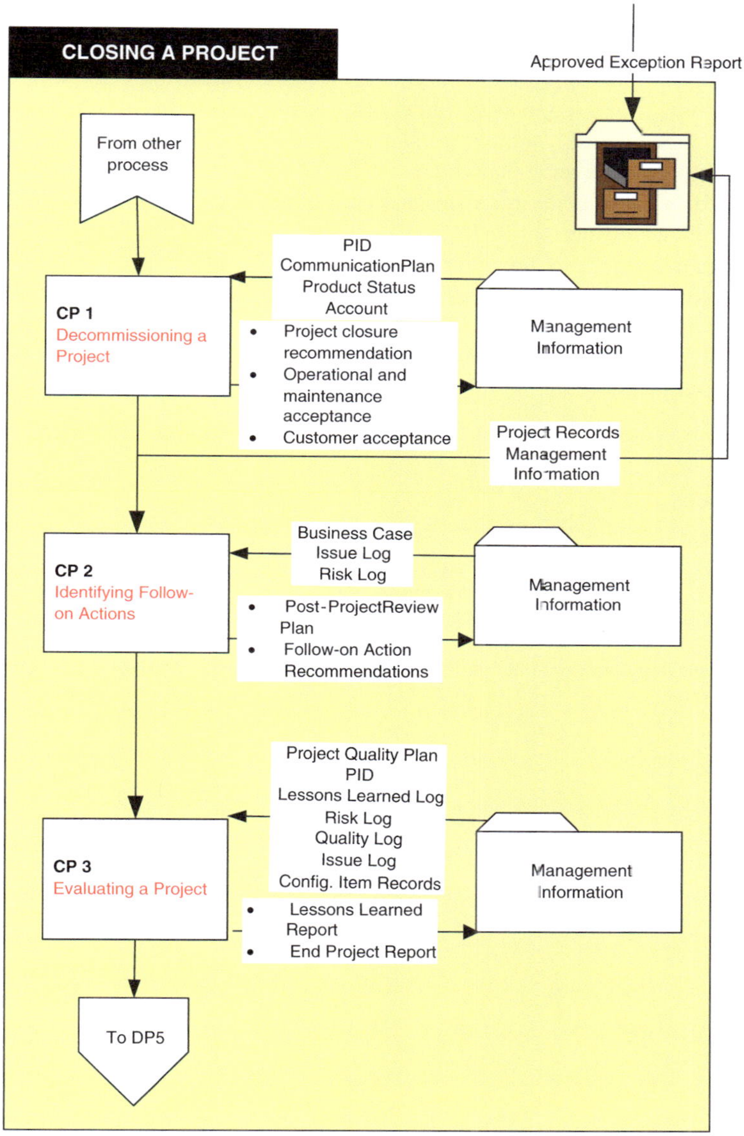

Fig. 8.7 PRINCE2® process closing a project

Additionally this citation is a valuable resource with further links on the subjects project planning and management. To summarize the most important requirements for a good plan it can be said that the plan must contain (obvious but worthwhile mentioning) a chronological listing of Products or Deliverables broken down into activities, actions and Tasks, especially applying the T^2M method for fine-grained planning of distributed activities, including the dependencies between the different Deliverables. A Plan requires a timeline and schedule, showing beginning and end points of the actions (aggressive but realistic); must state the resources required for each action or task, notably persons, person-days and money; must contain milestones to be achieved meanwhile.

To perform the overall organizational coordination, tool support for the project management according to PRINCE2® is required. The PRINCE2® standard already contains quite a few predefined checklists and templates to coordinate diverse activities but additional tools are required. These tools need to cover the support of the Planning and Tracking itself. Additionally a constant presentation of the current project status is preferable to allow early noticing of deviations and easy status communication.

References

1. Bentley, C.: PRINCE2 a Practical Handbook, 3rd edn. ButterworthHeinemann, Oxford (2009)
2. Chemuturi, M.K., Cagley, T.M.J.: Mastering Software Project Management: Best Practices, Tools and Techniques. J. Ross Publishing, Plantation (2010)
3. Gido, Jack: An Introduction to Project Planning. Industrial, New York (1985)
4. OGC: Managing Successful Projects with PRINCE2, 4th edn. The Stationery Office Ltd, Norwich (2009)

Glossary

Deliverable A Deliverable is the actual outcome of a Task. Like in Stage-Gate® this has to be something tangible like a document, a computer program, or a physical artefact. Spent money or wasted time may be the main outcome of some Tasks but do NOT count as Deliverables. An additional restriction of T^2M is that it forbids Deliverables that create overhead. There may never be a Deliverable that is created only for the sake of a Transition. A Deliverable should be created to advance the Project and not the process.

Each Deliverable has a specification, an implementation, and a rating. The specification describes the requirements and constraints of the Deliverable. The implementation is the resulting Deliverable, and the rating describes how the implementation matches the specification. In its most simple form this could be "accepted" or "rejected."

Fabless If a company develops a semiconductor product without owning a semiconductor fabrication line, the products are developed fabless.

IP Intellectual Property. The legal rights associated with inventions, artistic expressions, and other products of the imagination, in the form of patents, trademarks, service marks, trade names, trade secrets, and copyrights.

Operations The Operations efforts of a business valuate products, services, and assets marketed by a business. Therefore operations take the results of an innovation project through the life cycle of the product. It can be distinguished from a project because it is ongoing and repetitive.

Product As [Wik05i] defines, a Product is anything that can be offered to a market that might satisfy a want or need of a potential customer. It is the complete bundle of benefits or satisfactions that buyers perceive they will obtain if they purchase the product. It is the sum of all physical, psychological, symbolic, and service attributes.

Gate Gates are the decision points in the lifetime of a Project where the Gatekeepers decide whether to start or continue funding a Project. The Gates are the key to distinguish between the potentially good and bad projects or products.

D. Ortloff et al., *MEMS Product Engineering*, DOI 10.1007/978-3-7091-0706-5,

At the Gate the Gatekeepers take tough decisions about the projects. The decisions are essential for the success of a new product strategy. Essentially they enable to focus scarce resources on the potentially best innovation projects, thus establishing value for money. The outcome of these funnel decisions is either to Go, Kill, Hold, or Recycle a Project.

Gates consist of three main elements which are the required deliverables, a set of evaluation criteria, and the outputs. The required deliverables document the advancement of the project. They are based on a predefined catalogue per Stage given by the engineering method and are agreed between the Gatekeepers and the project management in the preceding Gate. The evaluation criteria are used to assess the project. They typically consist of knockout questions like strategic fit used to weed out not fitting products and a set of should-meet criteria scoring the project. The overall scoring of a project is used for prioritizing the different projects. The outputs of a Gate, in the positive case, consist of the funding decision, an approved action plan for the next Stage and a list of Deliverables and criteria for the next Gate.

Gatekeepers The Gates are executed by the Gatekeepers of the process. The Gatekeepers are a team of medium or senior leaders taking Go/Hold/Recycle/Kill decisions at the Gates. They are from different functional areas and have to be entitled to commit resources to a project. They use a pre-set list of criteria and rules for the Gates to prevent promoting "personal pet projects." By those rules and criteria they are bound so that they cannot play favorites. The first three Gates of the process can be performed by lower level management, e.g., one design, one process and one modeling group leader, the CTO, and a marketing representative. Later Gates will be performed by the VPs of the different business areas, including financial and other involved business groups.

Product Engineering (PE) Product Engineering is the discipline of driving and managing product innovation efforts following a predefined methodology. It considers the whole life cycle of a Product from its idea over the various development Stages to a marketable Product. A more detailed definition is given in Sect. 1.2.

Product Engineering Framework (PEF) The discipline Product Engineering requires tool and process components to perform the related tasks. The Product Engineering Framework contains such a tool framework supporting the efforts to drive a product idea to a marketable product. For the definition of the different steps it uses an engineering methodology defined in the Product Engineering Method (PEM) integrated with the Quality Assurance Framework.

Product Engineering Method (PEM) Defines a methodology, consisting mainly of processes and procedures, driving product innovation. The components of the methodology consist of business processes and subprocesses defining the sequence of actions to engineer a product idea into a marketable product.

Program A Program is a collection of Projects or Product innovations to address a strategic objective of the overall strategy of a business.

Project A Project is a temporary endeavor to create a product, a service, or any other result like a cultural change. Therefore a project ends when its objectives are achieved, e.g., a product is launched. It is typically constrained by scope goals, time goals, and cost goals. For Product Engineering all efforts to evaluate a product idea and engineer a Product are considered to be a development or innovation project. Therefore a Project is the innovation phase (from the idea to a marketable product) of the product life cycle and hands over the innovation results to Operations (defined above).

Project Management (PM) are the activities to guide a project via the application of knowledge, skills, techniques, and tools typically performed by the Project Manager.

Quality Assurance Framework (QAF) The Quality Assurance Framework defines necessary quality assurance activities. It comprises of, e.g., review procedures and quality audit procedures ensuring that product deliverables are developed until a certain quality level is reached and that they are developed in accordance with a predefined quality assurance model.

Stage Each Stage of an innovation Project contains a set of defined concurrent activities, incorporating industry best practices. Activities during each Stage are executed in parallel to enhance speed to market and are focused on the Deliverables of the development Stage. To foster the focusing onto the Deliverables a Deliverable-oriented, not activity-oriented, planning process is used to establish the Stage plans. The speed to market is further enhanced by the use of multidisciplinary teams all working towards the same goals and on the same set of Deliverables required for the next Gate. The Stages are primarily focused to produce high quality Deliverables to pass the next Gate only. Typical Stages have growing sizes during the lifetime of a Project where each later Stage is bigger than a previous one in terms of commitment, effort, money, etc. On the other hand the risks and the uncertainties of the Project have to decrease to enable successful assessment during the Gates.

Stakeholder A corporate Stakeholder is a party who affects, or can be affected by, the company's actions. The stakeholder concept was first used in a 1963 internal memorandum at the Stanford Research institute. It defined stakeholders as "those groups without whose support the organization would cease to exist." The theory was later developed and championed by R. Edward Freeman (Freeman, R.E.: Strategic Management: A Stakeholder Approach. Pitman, Boston (1984). ISBN:0-273-01913-9). Since then it has gained wide acceptance in business practice and in theorizing relating to strategic management, corporate governance, business purpose, and corporate social responsibility (CSR). In principle it refers to everybody or every role having an interest in something.

Task In T^2M tasks represent activities that implement deliverables or to put it in less abstract words: tasks are activities whose results are called Deliverables, e.g., the Deliverable of a task "Market Analysis" would be a "Market Analysis Report." Therefore the most basic Task consists of a name that describes the overall activity, a list of Deliverables that have to be implemented, and last but not least the budget that has been assigned to perform the Task.

Transition Transitions in the T^2M serve as a synchronization point between Tasks. Here the results (Deliverables) of the preceding Task are collected and handed over to the subsequent Tasks. A Transition is activated if all required Deliverables are accepted. After activation of a Transition all preceding Tasks are closed and all subsequent Tasks are activated. Therefore a Transition is similar to a Gate, although much less formal in nature.

Index

D. Ortloff et al., *MEMS Product Engineering*, DOI 10.1007/978-3-7091-0706-5,

MIX
Papier aus verantwortungsvollen Quellen
Paper from responsible sources
FSC® C105338

If you have any concerns about our products,
you can contact us on
ProductSafety@springernature.com

In case Publisher is established outside the EU,
the EU authorized representative is:
Springer Nature Customer Service Center GmbH
Europaplatz 3, 69115 Heidelberg, Germany

Printed by Libri Plureos GmbH
in Hamburg, Germany